Molecular Docking in Biomedical Engineering and Computational Chemistry

Edited by Rohit Bhatia

Published in London, United Kingdom

Molecular Docking in Biomedical Engineering and Computational Chemistry
http://dx.doi.org/10.5772/intechopen.1009057
Edited by Rohit Bhatia

Contributors
Aastha Rana, Amr Tayel, Bandenahalli Siddalingappa Krishna, Bilal Shaker, Gagandeep Singh, Habiba Zeb, Haytham O. Tawfik, Ishaq N. Khan, Khaled Barakat, M. Modather Taha, Monochura Saha, Rohit Bhatia, Shubham Yadav, Tamer Shoeib

First published in London, United Kingdom, 2026 by IntechOpen
IntechOpen is the global imprint of INTECHOPEN LIMITED, registered in England and Wales, registration number: 11086078, 167-169 Great Portland Street, London, W1W 5PF, United Kingdom

For EU product safety concerns: IN TECH d.o.o., Prolaz Marije Krucifikse Kozulić 3, 51000 Rijeka, Croatia, info@intechopen.com or visit our website at intechopen.com.

British Library Cataloguing-in-Publication Data
A catalogue record for this book is available from the British Library

Molecular Docking in Biomedical Engineering and Computational Chemistry
Edited by Rohit Bhatia
p. cm.

This title is part of the Biomedical Engineering Book Series, Volume 37
Topic: Bioinformatics and Medical Informatics
Series Editor: Robert Koprowski
Topic Editor: Slawomir Wilczynski

Print ISBN 978-0-85014-754-4
Online ISBN 978-0-85014-753-7
eBook (PDF) ISBN 978-0-85014-755-1
ISSN 2631-5343

If disposing of this product, please recycle the paper responsibly.

IntechOpen Book Series

Biomedical Engineering

Volume 37

Aims and Scope of the Series

Biomedical Engineering is one of the fastest-growing interdisciplinary branches of science and industry. The combination of electronics and computer science with biology and medicine has improved patient diagnosis, reduced rehabilitation time, and helped to facilitate a better quality of life. Nowadays, all medical imaging devices, medical instruments, or new laboratory techniques result from the cooperation of specialists in various fields. The series of Biomedical Engineering books covers such areas of knowledge as chemistry, physics, electronics, medicine, and biology. This series is intended for doctors, engineers, and scientists involved in biomedical engineering or those wanting to start working in this field.

Meet the Series Editor

Robert Koprowski, MD (1997), Ph.D. (2003), Habilitation (2015), is an employee of the University of Silesia, Poland, Institute of Computer Science, Department of Biomedical Computer Systems. For 20 years, he has studied the analysis and processing of biomedical images, emphasizing the full automation of measurement for a large inter-individual variability of patients. Dr. Koprowski has authored more than a hundred research papers with dozens in impact factor (IF) journals and has authored or co-authored six books. Additionally, he is the author of several national and international patents in the field of biomedical devices and imaging. Since 2011, he has been a reviewer of grants and projects (including EU projects) in biomedical engineering.

Meet the Volume Editor

Dr. Rohit Bhatia has 13 years of experience in academics, administration, and research. He specializes in Medicinal Chemistry, Computer-Aided Drug Design, and Bioanalysis. His areas of interest include anticancer drug discovery and paper-based microfluidics. He has published more than 150 research and review articles in various international and national journals. He has published more than 12 book chapters. He has delivered more than 20 guest lectures as a speaker on various National and International platforms. He has qualified for the GPAT exam four times. He has guided 32 postgraduate students and 6 Ph. D students. He is a potential reviewer for various internationally reputed journals. He served as the brand ambassador for Bentham Science in 2019 and 2024. He has also presented his research at various national and international conferences. The Government of India has granted him 12 patents. Recently, he received a grant of 20 lakh from MSME, India, to develop a point-of-care device for detecting glucose in human saliva. He has been awarded the Young Scientist Award twice on international platforms. He has also authored six books.

Contents

Preface

The rapid advancement of computational tools has profoundly transformed the landscape of modern biomedical research and drug discovery. Among these tools, molecular docking has emerged as a cornerstone technique, enabling researchers to predict and analyze biomolecular interactions with remarkable efficiency and accuracy. The book "Molecular Docking in Biomedical Engineering and Computational Chemistry" is conceived to provide a comprehensive, structured, and application-oriented understanding of molecular docking and its pivotal role in contemporary biomedical and chemical research.

This book bridges fundamental concepts with practical applications, addressing the needs of students, researchers, and professionals working at the interface of biomedical engineering, pharmaceutical sciences, and computational chemistry. It introduces the theoretical foundations of molecular recognition, binding energetics, and scoring functions, while progressively guiding the reader through docking algorithms, software tools, validation strategies, and interpretation of docking results. Special emphasis has been placed on real-world biomedical applications, including drug–target interactions, enzyme inhibition, protein–protein interactions, and structure-based drug design.

Recognizing the interdisciplinary nature of the subject, the chapters are designed to integrate principles from chemistry, biology, mathematics, and computer science in a coherent and accessible manner. Case studies, illustrative figures, and examples from current literature have been incorporated to enhance conceptual clarity and practical relevance. The book also highlights recent advancements, including flexible docking, virtual screening, and the integration of molecular docking with molecular dynamics and artificial intelligence.

This book is intended to serve as a valuable resource for undergraduate and postgraduate students, doctoral researchers, academicians, and industry professionals. It may be used both as a teaching companion and as a reference for research and innovation in computational drug discovery and biomedical engineering.

The authors hope that this book will not only strengthen foundational knowledge but also inspire readers to explore and apply molecular docking techniques in solving complex biomedical challenges. Constructive feedback from readers is warmly welcomed for the continual improvement of future editions.

Dr. Rohit Bhatia
Associate Professor and Head,
Department of Pharmaceutical Chemistry,
Chitkara College of Pharmacy, Chitkara University,
Rajpura, Punjab, India

Chapter 1

Beyond Docking: Intelligent Modeling and the New Landscape of Drug-Protein Interactions

M. Modather Taha, Amr Tayel, Haytham O. Tawfik and Tamer Shoeib

Abstract

Molecular docking is a central technique in computer-aided drug design (CADD), offering a way to predict and visualize how potential drugs interact with their protein targets. By modeling binding poses, interaction patterns, and binding strengths, docking helps researchers move beyond trial-and-error approaches toward more rational drug design. Its value is not limited to designing new molecules; it also plays a key role in drug repurposing, where existing compounds can be redirected to treat new diseases. To ensure reliable results, benchmarking strategies have been developed, allowing docking methods to be tested and validated across large datasets. At the same time, protein engineering has enhanced the accuracy of docking by improving the quality of target structures and revealing new binding opportunities. More recently, artificial intelligence (AI) has pushed the field forward by refining pose predictions, improving scoring systems, and uncovering new repurposing opportunities. Taken together, these advances show that molecular docking has evolved from a static predictive tool into a dynamic, integrated framework that combines structural biology, computational methods, and AI to drive faster and more effective drug discovery.

Keywords: molecular docking, drug repurposing, protein engineering, benchmarking, artificial intelligence

1. Introduction: The Simple Truth - Drugs act by binding to proteins

Behind every pill of medication we take, many complex mechanisms must take place inside our body to produce the final therapeutic effect. We are often concerned with the outcome without asking how it actually happens. From a classical perspective, it was always acceptable and considered normal to discover a drug that had a great clinical outcome without fully understanding how it interacted with the body. Over the decades, scientists have realized the significant shift that pharmaceutical companies have undergone to advance the drug development process to a more rational and mechanistic level. This shift involved moving from a reliance on natural drugs to synthetic chemistry and the chemical optimization of existing drugs. Consequently, there were strict requirements for this transition, as pharmaceutical companies could no longer afford to synthesize or develop drug candidates blindly, wasting time and money without knowing how drugs truly interact with the body. Therefore, they relied on the core concept of pharmacodynamics, which simply states that any drug must interact with at least a single protein to exert a pharmacological effect and ultimately produce a clinical outcome [1].

Accordingly, drug–protein interactions need to be studied in greater depth to assess newly synthesized drug candidates or to provide insights for the optimization of existing drugs. In fact, identifying the types of drug–protein interactions (such as hydrogen bonds, ionic bonds, and van der Waals forces) and evaluating their strength and specificity opens the door to extensive modification strategies for investigational compounds. Additionally, studying the thermodynamic properties of these interactions can be highly beneficial for optimizing drug candidates or selecting the most promising ones for further development.

However, such detailed investigations were practically unfeasible in the decades prior to the 2000s. In the late 1990s, computer-aided drug design (CADD) began to emerge, introducing a variety of virtual techniques capable of aiding drug discovery and design. This marked a major milestone in the drug development process. These virtual methods were collectively referred to as in silico approaches. Among the most powerful and indispensable techniques introduced by the CADD approach for analyzing drug–protein interactions was molecular docking [2].

1.1 Molecular docking

The power of visualization in uncovering more fundamental and deeper insights has led scientists to rely heavily on the CADD approach, particularly on molecular docking. Molecular docking is a CADD technique that places a predictive lens within the computer to forecast a glimpse of the molecular events inside the human body that drive biological pathways. In general terms, docking is a rapid computational tool that assesses the binding pattern between a pair of chemical entities. These pairs may include ligand–protein, protein–protein, or ligand–nucleic acid interactions [3]. From a drug discovery perspective, docking is used to predict how well a compound of interest fits onto a specified target, allowing for the virtual assessment of its binding, as shown in **Figure 1**. From another point of view, it is a computational technique capable of investigating and visualizing the energetically

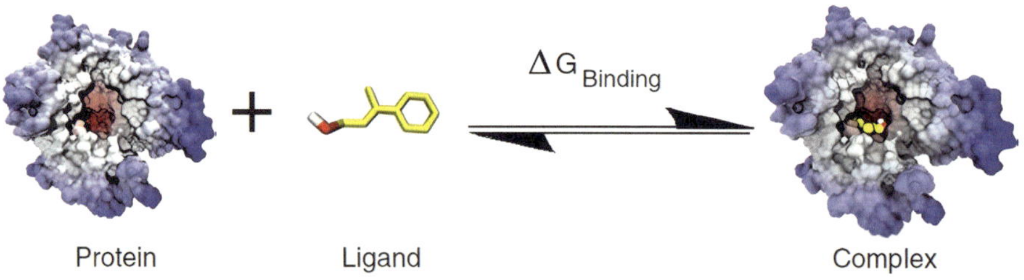

Figure 1.
Basic docking protocol showing how the ligand fits into the protein [6].

favorable conformations or poses in which a ligand may adopt within the binding pocket of a given target [4]. Moreover, it identifies the key interactions responsible for driving the binding affinity between the two molecules. Importantly, it quantifies the quality of each pose by estimating its binding energy. This virtual technique uses scoring functions and sampling algorithms to perform docking analysis. Several software packages perform docking using different algorithms, including Glide, MOE, GOLD, and AutoDock Vina.

The results of molecular docking yield two main outcomes: one quantifiable and the other qualitative. These include the docking score (which is not an absolute value, as it varies depending on the software used), the predicted interactions with the target protein, and the energetically favorable poses of ligand binding [5, 6]. In practice, to perform any docking protocol using a software package, two main inputs are required: the 2D structure of the ligand of interest (or a collection of ligands from a database) and the 3D crystal structure of the target protein, which can be accessed from the Protein Data Bank (PDB) [7].

This crystal structure is typically obtained through experimental techniques such as X-ray crystallography, NMR spectroscopy, or cryo-electron microscopy (cryo-EM). The quality of the obtained crystal structure is a critical factor in ensuring the reliability of the docking protocol [6, 8].

In the scope of industrial applications, docking is not only important for designing novel structures – a method known as structure-based drug design – but also has broad applications that establish it as a critical stage in any drug development pipeline. Examples include the lead optimization approach, which involves modifying potential ligands or drug candidates to enhance their pharmacokinetic profile, improve potency, and reduce observed in vivo side effects. Another key application is drug repurposing, which will be discussed in the following section [8, 9].

There are two main types of docking: rigid docking and flexible docking. Rigid docking does not account for conformational flexibility, as it predicts interactions between two static entities. Although it is computationally fast, it lacks sufficient accuracy and is therefore mainly used for rapid high-throughput screening. On the other hand, flexible docking allows for ligand flexibility, enabling the exploration of various conformational spaces. It provides more accurate and reliable results, as the ligand typically samples multiple conformations until it reaches the most stable one. Therefore, it will be thermodynamically capable of interacting with the target protein in a meaningful and biologically relevant manner [5].

1.2 Drug repurposing

Logically, it is not possible to control the interaction map of a drug within the body with 100% precision; therefore, drugs often bind to unintended protein targets in addition to their primary target of interest. This reality led scientists to consider how off-target activities or side effects could be utilized beneficially [9]. One of the most notable examples of this concept emerged from a famous case of serendipity in the pharmaceutical industry, which led to the discovery of Viagra. Originally developed as an antihypertensive agent, Viagra was found during clinical trials to have a significant side effect: a potent vasodilatory effect, which notably improved penile blood perfusion. As a result, its therapeutic focus was shifted, and it was repurposed to treat erectile dysfunction as its primary indication. Numerous similar serendipitous findings have resulted in the discovery of drugs for indications different from those originally intended [10]. This phenomenon laid the foundation for a pivotal strategy in the drug discovery process known as drug repurposing. Drug repurposing involves reorienting market-approved drugs or unapproved drug candidates toward new therapeutic indications. It is considered a cost-effective, faster, and lower-risk alternative to traditional de novo drug discovery [11, 12].

Recently, the COVID-19 pandemic has demonstrated to the pharmaceutical industry the significant advantages of drug repurposing, as many marketed drugs were re-investigated for COVID-19 treatment following the FDA's emergency use authorization of new therapies [13, 14]. The primary advantage of this strategy lies in its ability to bypass the long and tedious steps typically required in drug development, such as in silico studies, pharmacokinetic testing, and safety profiling. Repurposing studies often aim to identify a drug's mechanism of action by investigating chemical structure similarities between compounds or by screening existing therapies against new targets to discover alternative indications [13]. The rapid expansion of biomedical databases, including those containing RNA expression signatures, pharmacological binding profiles, bioavailability data, and biobanks with clinical, molecular, and genomic data, has significantly fueled the development of computational methods in drug repurposing. The complex relationships among biomedical entities, such as genes, drugs, diseases, and adverse reactions, are central to repurposing studies, with various methodologies emphasizing different types of data depending on the objective and design of the analysis [11, 15].

2. Beyond basic docking

As discussed, molecular docking has become an indispensable technique in CADD, structural biology, and indeed across the entire workflow of modern drug design. Moving from theoretical foundations into practical applications, there exists a wide range of pharmaceutical pipelines where docking contributes substantially, depending on the case study and the objectives pursued in the drug discovery project.

2.1 Applications of molecular docking

One of the most traditional and well-established applications of docking is virtual screening, which serves as a cost-effective and rapid in silico strategy to identify new potential drug candidates. With the continuous daily expansion of chemical space – driven by the accumulation of millions of novel molecules from diverse chemical databases – there arises a pressing need for strategies capable of quickly assessing these molecules. Virtual screening through docking fulfills this role. The ultimate goal is to computationally screen enormous chemical collections, whether they are commercial libraries, lead-like compound repositories, or fragment-based libraries, against a well-prepared and validated protein target of interest [14]. This allows researchers to prioritize a smaller number of promising "hit" compounds for subsequent optimization. As shown in **Figure 2**, one of the most popular approaches to screen active compounds for a specific target is pharmacophore-based virtual screening [15].

DOI: http://dx.doi.org/10.5772/intechopen.1013595

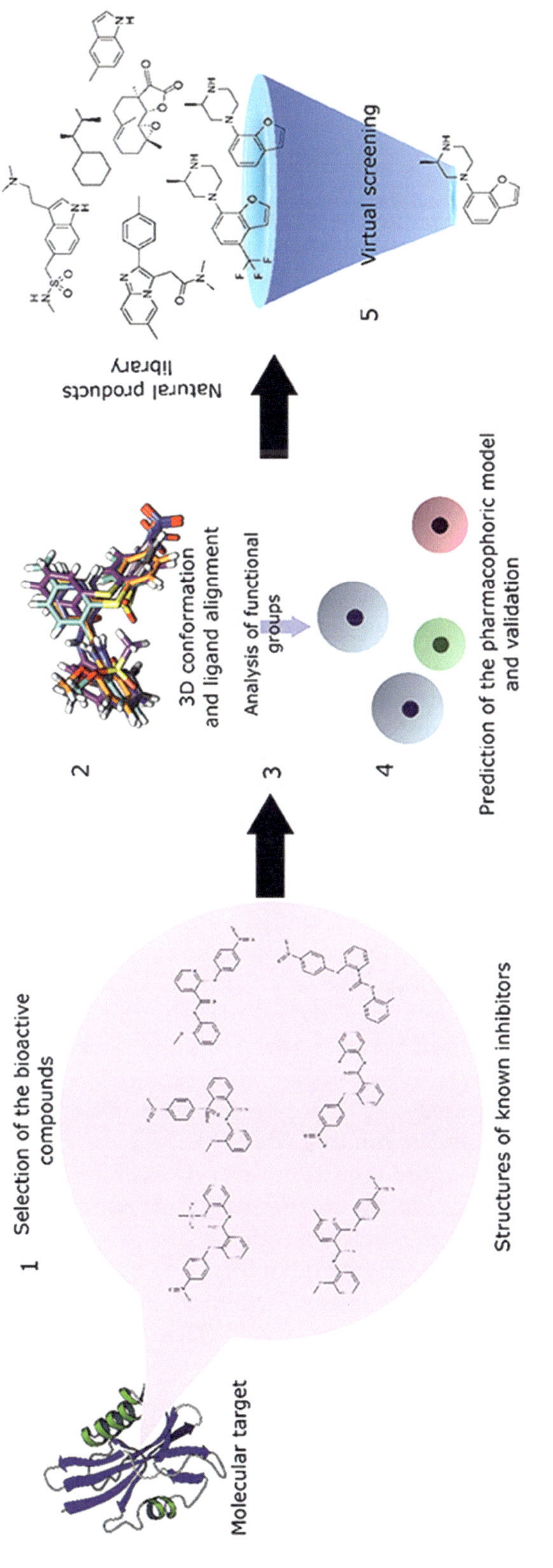

Figure 2.
An illustration of a virtual screening pipeline based on pharmacophore modeling [15].

In this context, docking provides approximate binding scores that can be used to filter out nonpromising molecules and retain candidates with favorable profiles. To enhance confidence in the results, it is common practice to employ multiple scoring functions in parallel and to calculate the root mean square deviation (RMSD) relative to a reference ligand. These technical steps act as safeguards, helping to reduce false positives and strengthen the reliability of the docking protocol [16]. Once the virtual screening is complete, the resulting hit compounds are either purchased or synthesized, allowing the workflow to proceed to experimental phases. This process emphasizes the essential role of docking as a rapid tool for exploring vast chemical spaces in industrial projects where efficiency and cost-effectiveness are critical [14, 16].

Another important application of docking is its extensive use in uncovering the chemical mechanisms underlying pharmacological effects. In practice, many drugs exhibit unintended but favorable outcomes during clinical trials, and pharmaceutical companies often advance these drugs toward approval without spending time investigating the molecular reasons behind the observed effects. Docking, however, can provide a convenient and powerful means to explore these phenomena. By simulating the interaction of a drug with a variety of possible protein targets, docking enables researchers to hypothesize initial mechanistic explanations and to identify which binding poses and interaction patterns might account for the observed pharmacological profile [17, 18].

The primary aim in such investigations is to determine the energetically favorable binding pose and the associated interaction map between the drug and the protein residues. Nevertheless, it is important to recognize that docking alone cannot provide a complete mechanistic understanding. The method yields approximate predictions and lacks the resolution to fully define the dynamic nature of drug–target interactions [17]. For this reason, docking outputs are often coupled with more advanced computational analyses, such as Molecular Mechanics Generalized Born Surface Area (MM-GBSA) calculations, which refine the predictions by delivering a more accurate assessment of binding energies [19]. Docking also represents the central technique underlying de novo drug design, a creative and highly innovative approach in structure-based drug discovery. De novo design involves generating entirely new ligands from scratch, guided by structural insights into the amino acid residues and the physical characteristics of the binding pocket. Before ligand design begins, the molecular surface, active pocket volume, and spatial arrangement of residues must be carefully analyzed to determine what initial geometry a ligand should adopt to complement the binding site [20].

Traditionally, chemists would approach this process by proposing sets of complementary ligand fragments tailored to individual residues in the pocket. These fragments were then linked together in a rational manner, with careful attention given to steric clashes, conformational flexibility, and the overall volume of the binding site. The newly constructed ligand would then be docked into the protein to evaluate how well it fits into the binding pocket. This was – and continues to be – an iterative process, repeated until a collection of ligands is identified that adopt favorable, energetically stable binding poses. Although this approach was first introduced decades ago, de novo design remains one of the most imaginative and enduring strategies in modern drug discovery, displaying docking's irreplaceable role in the design of novel ligands [21]. A great example is Ribociclib (Kisqali), a CDK4/6 inhibitor used for HR-positive, HER2-negative breast cancer. Scientists studied the ATP-binding pocket of CDK6 and, using fragment-based design with docking, built and refined small pieces into a drug-like molecule. Step by step, these fragments were linked and optimized until they created a highly selective and potent inhibitor. Ribociclib shows how docking-guided design can turn structural insights into real, life-saving medicine [22].

2.2 Postrefinement analyses to overcome docking limitations

Despite its importance, docking is not without limitations. It is widely acknowledged that docking provides an approximate view of binding and cannot be expected to perfectly correlate with experimental assay outcomes. The analogy often used is that docking resembles a static photograph: it captures an image of the interaction at a single moment in time, providing useful insights, but it does not show the full movie of how binding actually occurs in a dynamic biological system. To bridge this gap, a number of postrefinement analyses have been developed to extend docking and overcome its inherent shortcomings.

One of the most common refinements is the MM-GBSA analysis. This method takes the binding poses obtained from docking protocols and subjects them to more rigorous free energy calculations. MM-GBSA simulates the dynamic binding process and produces more accurate estimates of binding energies by incorporating parameters that docking alone does not handle effectively, such as solvation effects, nonpolar interactions, and both enthalpic and entropic contributions [23]. The outcome is a fuller thermodynamic profile of the interaction between the ligand and the target. Although MM-GBSA is not typically used in large-scale virtual screening due to its higher computational cost, it plays a crucial role in refining docking results and validating hits before committing to experimental studies [24].

Another powerful refinement approach is the protein–ligand interaction fingerprint (PLIF). This technique originates from the recognition that every interaction between a ligand and a protein has a unique identity that can be captured and encoded for systematic study. In practice, docking results from different ligand sets are collected, and instead of depending solely on docking scores, researchers encode the specific interactions into binary fingerprints. Each element of the fingerprint corresponds to a particular type of interaction with specific protein residues. By effectively transforming the 3D interaction map into a simplified and comparable digital format, PLIF enables researchers to compare, analyze, and store interaction data in a highly efficient manner. PLIF becomes especially valuable when integrated with machine learning approaches. By training models on fingerprints derived from docking simulations, predictive algorithms can be developed to recognize recurring binding patterns and binding motifs. This approach is particularly effective in targets with relatively straightforward binding modes, where original ligands can serve as templates. Docking thus provides the structural input, while PLIF translates the interaction data into a form that can power advanced computational analyses and predictive modeling [25].

Another important refinement is solvent analysis, which focuses on the critical role of water molecules in molecular recognition and complex formation. In realistic biological systems, water often mediates binding by either being displaced to permit direct ligand–protein contact or by remaining in place to bridge interactions. The thermodynamic contribution of this solvent reorganization is significant and directly impacts ligand binding affinity through the phenomenon of desolvation energy [26].

Ignoring solvent effects can, therefore, lead to discrepancies between computational predictions and experimental results. To address this, explicit molecular dynamics simulations, such as those performed using GROMACS, are often applied to study solvent effects. These simulations provide insights into how water molecules behave during ligand binding and allow for more accurate modeling of the thermodynamic landscape. By incorporating solvent effects into docking refinements, the predictions become more realistic and are better correlated with experimental data [24, 27].

A further advanced refinement strategy involves quantum mechanics/molecular mechanics (QM/MM) calculations. A common limitation of many docking programs is the assignment of fixed partial charges to ligands. This simplification can distort the actual distribution of electronic density, particularly for flexible ligands with numerous rotatable bonds. Such inaccuracies propagate into errors in polarization energies and, ultimately, into the reliability of docking predictions. QM/MM methods overcome this by recalculating the electronic structure of ligands dynamically within the protein environment [28]. Programs such as QSite employ self-consistent field theory to determine partial charge distributions with much greater accuracy than fixed-parameter docking approaches. The integration of QM/MM into docking refinements produces more reliable binding poses and energy calculations, effectively bridging the gap between classical docking and quantum-level simulations [29].

In conclusion, molecular docking should be viewed as a foundational tool that supports and interconnects with multiple layers of refinement. Its direct applications in virtual screening, mechanism elucidation, and de novo drug design demonstrate its breadth, while postdocking refinements such as MM-GBSA, PLIF, solvent analysis, and QM/MM corrections enhance its predictive power. Collectively, these strategies transform docking from a static approximation into a dynamic, thermodynamically rigorous, and quantum-aware predictive framework. Thus, docking not only remains central to drug discovery pipelines but also continues to evolve as computational methodologies advance [30].

2.3 Benchmarking

Molecular docking can be viewed as a puzzle-solving process in which computational tools are used to assess how effectively a drug molecule fits into the binding site of a protein. To validate the performance of these tools, researchers rely on advanced benchmarking strategies that evaluate their accuracy in predicting binding modes, binding affinities, and their ability to identify promising drug candidates from large molecular libraries. Recent improvements in benchmarking approaches have significantly refined these evaluations, contributing to more efficient drug discovery.

One major advancement is enhanced pose prediction, which measures whether docking methods can accurately reproduce the 3D orientation of a ligand within a protein. High-quality experimental datasets, such as the PDB, are often used for this purpose. For instance, a 2021 study assessed the performance of tools like Glide and AutoDock Vina against 193 SARS-CoV-2 protease complexes, using RMSD as the success metric, with values below 2 Å considered ideal. More recently, AlphaFold2-predicted structures have expanded this testing to include computationally generated protein models, allowing evaluation under conditions of structural uncertainty [31].

Virtual screening benchmarks have also progressed, with datasets such as DUD-E and the more recent LIT-PCBA (2020) providing realistic challenges by combining true binders with decoy molecules. Performance is then measured through enrichment metrics such as ROC-AUC and enrichment factors. A further development came in 2025 with the lsd.docking.org database, which included 6.3 billion docked molecules across 11 protein targets, offering a large-scale resource for testing both docking algorithms and machine learning models in virtual screening. Together, these advanced benchmarking approaches – spanning pose prediction, affinity evaluation, and large-scale screening – are strengthening the reliability of molecular docking and accelerating the pathway to effective drug discovery [32].

3. Protein engineering

Proteins are highly complex macromolecules with dynamic properties and distinct functional domains. One of the major limitations of molecular docking lies in treating proteins as rigid, immobile structures, which oversimplifies their biological reality. While additional layers of analysis can partially address this, challenges often arise when the protein itself becomes the weak link in the design workflow. Focusing extensively on generating new hits through de novo design or virtual screening, without carefully considering the accuracy and behavior of the target protein, can lead to misleading outcomes in drug discovery.

Having access to high-resolution crystal structures of the target protein is a considerable advantage, yet in many cases, such structures are unavailable or only incomplete models exist. To accurately reproduce the binding profile of ligand–protein interactions, both molecules must be computationally optimized to reflect their biological dynamics. Thus, the innovation in drug discovery is not limited to designing potential drug candidates but also extends to ensuring the target protein is represented at the highest possible quality – free of misfolding or structural artifacts – and that the binding pocket is precisely defined, accounting for the mechanical roles of its subpockets and loops within the active site.

A clear example is Imatinib (Gleevec), the first tyrosine kinase inhibitor approved for chronic myeloid leukemia. Its success came from looking closely at the target protein, BCR-ABL kinase, and realizing it could shift between active and inactive shapes. Early models treated the protein as rigid and missed the key "DFG-out" shape needed for binding. Once this hidden form was solved, scientists designed Imatinib to lock the protein in its inactive state. The drug's story shows that progress in drug discovery relies not just on designing new molecules, but also on truly understanding the proteins they target [6].

3.1 Protein engineering integration with docking

Protein engineering has become integral to refining molecular docking approaches by enabling the precise tailoring of drug targets, ultimately supporting more efficient drug discovery. Within structure-based drug design, techniques such as site-directed mutagenesis and directed evolution allow researchers to alter proteins in ways that enhance stability, solubility, and conformational accuracy – factors essential for reliable docking simulations, as shown in **Figure 3**. For example, engineering proteins to adopt biologically relevant conformations provides high-resolution structural templates that reduce uncertainty in ligand-binding predictions. This extends further to modifying proteins in ways that mimic physiological conditions, including posttranslational modifications or allosteric shifts, thereby improving the accuracy of virtual screening and lead identification for complex diseases. Moreover, by reconfiguring proteins to reveal hidden binding pockets or stabilize otherwise flexible regions, protein engineering increases the accessibility of targets to computational modeling and experimental validation, strengthening their role in modern drug design [6].

One of the major advantages of protein engineering in molecular docking is its ability to enhance docking accuracy by generating engineered proteins that overcome the shortcomings of conventional scoring functions and pose predictions. Traditional docking approaches often face challenges arising from protein flexibility and solvent influences, whereas engineered variants can be tailored to restrict conformational ensembles, producing more reliable ligand–protein interaction models [33]. For instance, refining scoring functions with correction factors derived from engineered protein–ligand complexes has been shown to improve docking and screening performance by better reflecting quantum mechanical influences and entropic effects. Likewise, integrating semiempirical quantum mechanical methods into docking workflows with engineered proteins enhances pose refinement, leading to higher success rates in both redocking trials and virtual screening studies. More advanced strategies, including the application of Markov State Models (MSM) built from engineered protein simulations, further improve structural reliability, surpassing conventional AlphaFold predictions in docking accuracy. Collectively, these improvements strengthen binding affinity estimations and streamline hit-to-lead development by reducing false positives in large-scale screening campaigns [33, 34].

Moreover, protein engineering supports the creation of target variants designed for specific docking applications, including multitarget therapies and complex binding environments, as shown in **Figure 3**. By producing engineered proteins with modified binding pockets or altered affinities, researchers can better replicate diverse docking conditions, such as covalent interactions or allosteric regulation, both of which are essential for tackling drug resistance and polypharmacology. Platforms like AlphaDrug apply de novo design principles to generate protein-specific ligands, while dynamic docking strategies integrate engineered variants to capture flexible interactions under realistic conditions. This approach is especially useful in cases where binders are developed directly from target structures, as engineered variants can be optimized for scaffold-based docking, ensuring suitability for therapeutic contexts such as enzyme inhibition or receptor antagonism [35]. Altogether, incorporating protein engineering into docking workflows marks a significant advancement, strengthening the connection between computational modeling and experimental validation, and ultimately driving more personalized and efficient drug discovery [36].

4. Artificial intelligence in modern drug discovery

Over the past decade, artificial intelligence (AI) has emerged as a global trend and has become increasingly integrated into the pharmaceutical industry across multiple

DOI: http://dx.doi.org/10.5772/intechopen.1013595

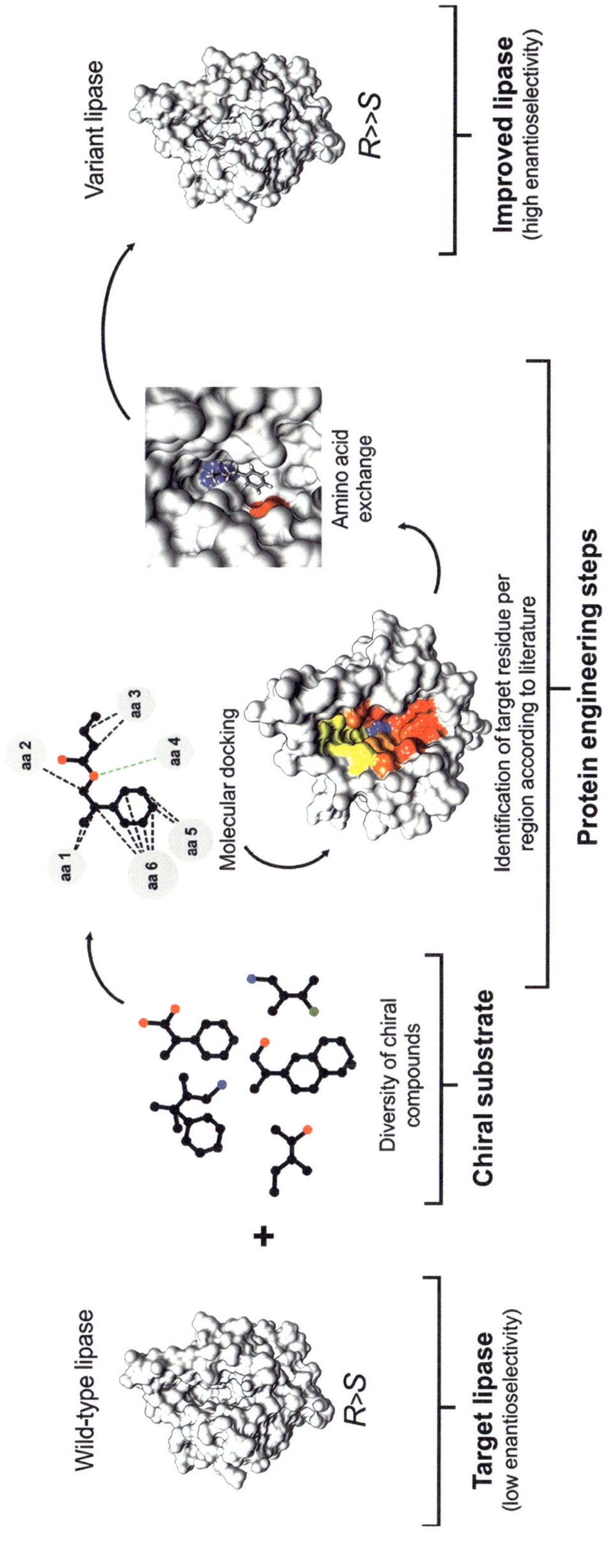

Figure 3.
Generating an improved variant of the target lipase using protein engineering [35].

domains. To briefly revisit the concept, AI is broadly categorized into supervised and unsupervised learning. Supervised learning encompasses three primary approaches: machine learning, deep learning, and reinforcement learning. Its core principle involves training a computational model to predict a dependent variable (the feature of interest) using independent variables as inputs, with performance depending heavily on the training process that enhances predictive accuracy. In contrast, unsupervised learning focuses on developing models that classify input data into meaningful clusters, often referred to as data segmentation. Both categories hold significant promise within the drug discovery pipeline, accelerating processes that traditionally require extensive time and resources. Looking ahead, AI-driven methodologies are expected to transform small-molecule design and shorten the timeline from discovery to clinical trials. Such advancements are critical for the pharmaceutical sector, which faces persistent delays in bringing new drugs to market. Nonetheless, concerns remain regarding how effectively AI can address essential safety and toxicity assessments. Still, one of the most notable achievements demonstrating AI's potential in this field is its successful contribution to the discovery of novel therapeutics [37]. A notable proof-of-concept for the potential of AI in transforming drug discovery is the work of Insilico Medicine, which successfully identified a TNIK inhibitor for the treatment of idiopathic pulmonary fibrosis. Remarkably, the entire process – from initial discovery to entry into clinical trials – was completed within only 18 months. This achievement highlights how AI-driven platforms can dramatically accelerate the traditionally lengthy drug development timeline and demonstrates the practical value of integrating AI into the pharmaceutical pipeline [38].

4.1 AI implementation in docking

AI is increasingly being utilized to enhance molecular docking through diverse approaches and computational methods. By leveraging large-scale chemical and biological databases, machine learning models can be trained to generate accurate predictions across different stages of the drug discovery pipeline. A crucial step in this integration is translating chemical and biological information into representations that computers can process [39]. Ligand structures, for example, can be encoded as SMILES strings, molecular fingerprints, graph-based networks, or 3D conformations, while protein data may be represented as amino acid sequences, transformed from PDB structures into 3D grids, or described through numerical pocket descriptors. Together, ligand–protein interactions can be modeled using graph–graph systems or 3D grid-based formats, enabling more precise computational simulations [39, 40].

One major application of AI in docking lies in pose prediction. Instead of relying solely on conventional scoring algorithms, AI models can rescore docking poses to improve accuracy, as shown in **Figure 4**. Reinforcement learning, using iterative trial-and-error training, allows models to refine binding orientation predictions until favorable poses are consistently identified. Deep learning methods, such as convolutional neural networks (CNNs), have also proven effective; a notable example is GNINA, an AI-augmented extension of AutoDock Vina, which employs CNN-based rescoring to improve pose prediction compared with traditional docking algorithms [41, 42]. Beyond pose prediction, AI has advanced the prediction of binding affinities by complementing or replacing classical scoring functions with

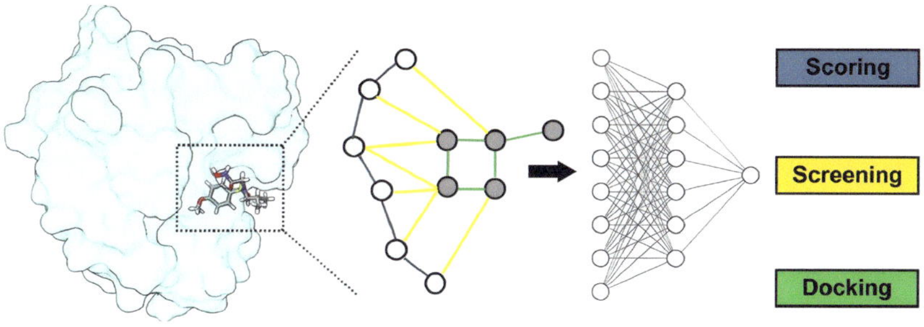

Figure 4.
Graph network representation of ligand-protein interaction using deep learning [41].

machine learning techniques, such as regression models, random forests, XGBoost, and deep neural networks, as demonstrated by platforms like DeepDock [43].

4.2 AI implementation in benchmarking

Machine learning methods can also advance benchmarking by improving the evaluation of docking accuracy and predictive performance. An example is DOCKSTRING, introduced in 2021, which provides large-scale datasets of more than 260,000 molecules across 58 targets for testing how effectively docking can predict binding affinities. These datasets use statistical measures, such as correlation coefficients, to compare predicted binding energies with experimental outcomes. Machine learning strengthens this process by refining scoring functions, as demonstrated with models like RF-Score and NNScore. In a 2022 study on antibiotic discovery, these methods improved docking accuracy, raising auROC scores from 0.48 to 0.63 44. Similarly, the LIT-PCBA dataset, released in 2020, introduced highly realistic benchmarking by combining active compounds with carefully selected decoys, offering a robust platform for testing virtual screening and scoring models. Together, such approaches make benchmarking more rigorous and enable docking tools to rank promising drug candidates with greater reliability [45, 46].

4.3 AI implementation in drug repurposing

Drug repurposing has rapidly evolved with the integration of AI, which has unlocked new opportunities by accelerating the identification of alternative therapeutic uses for existing drugs. AI-driven approaches enable the systematic mining of available biomedical knowledge and its alignment with diverse disease contexts. Several methods have been developed to achieve this. One major strategy is network-based repurposing, where AI models construct drug–target–disease interaction networks [47]. This is often implemented through graph neural networks, which predict associations by mapping disease–gene interactions as edges and drug–protein interactions as networks, then integrating the data to identify points of overlap. Another approach involves virtual screening of approved drugs against new targets, as demonstrated by DiffDock, which rescores docking poses to propose novel drug–target matches and potential indications [48].

Among the most advanced platforms in this area is CANDO, an AI-based drug repurposing system. It encodes input drugs into computational representations and virtually docks them across the entire proteome to generate interaction signatures [49]. These signatures are then compared, with similarity thresholds applied to identify drugs that may share therapeutic effects. By detecting such overlaps, the platform highlights opportunities for existing drugs to be repurposed or optimized as new leads. Overall, AI-powered repurposing has demonstrated efficiency in identifying novel therapeutic candidates, significantly reducing the time required in the drug discovery pipeline. Moreover, it represents a cost-effective alternative to de novo design or conventional virtual screening, since the candidate drugs already have established safety and toxicity profiles [50].

5. Conclusion

Molecular docking has become an essential tool in modern drug design, combining the speed of computational methods with valuable biological insights. Its strength lies in modeling how potential drugs fit and bind to their targets, which makes it useful not only for designing new molecules but also for tasks like virtual screening and drug repurposing. However, docking on its own provides only a partial picture of the complex and dynamic nature of biological interactions. To close this gap, researchers now combine docking with refinement techniques and structural innovations that make predictions more reliable. Benchmarking studies have added further confidence by testing docking methods across standardized datasets. At the same time, protein engineering has improved accuracy by producing better-quality target structures and opening new binding opportunities, although engineered proteins may not always represent native dynamics and can be costly and technically demanding. Drug repurposing has expanded the scope of docking by identifying new uses for existing compounds, but repurposed drugs often face challenges such as weak efficacy, off-target effects, or intellectual property barriers. The most recent leap has come from AI, which is reshaping docking through smarter scoring, improved pose prediction, and advanced repurposing strategies. Yet, AI approaches remain dependent on large, high-quality datasets, often act as "black boxes," and require careful validation before clinical translation. Taken together, these advances show how docking is steadily moving beyond its old limitations. It is no longer just a static tool but an evolving, adaptable, and powerful approach to drug discovery, yet still with challenges that need to be addressed.

Conflict of interest

The authors declare no conflict of interest.

Author details

M. Modather Taha[1], Amr Tayel[1,2], Haytham O. Tawfik[3] and Tamer Shoeib[2*]

1 Department of Pharmaceutical Chemistry, Faculty of Pharmacy, Alsalam University in Egypt, Tanta, Egypt

2 Department of Chemistry, American University in Cairo, New Cairo, Egypt

3 Department of Pharmaceutical Chemistry, Faculty of Pharmacy, Tanta University, Tanta, Egypt

*Address all correspondence to: t.shoeib@aucegypt.edu

References

[1] Schramm ME, Hu MY. Perspective: The evolution of R&D conduct in the pharmaceutical industry. Journal of Product Innovation Management. 2013;**30**:203–213

[2] Masood I, Ahmad M. Evolution of marketing techniques, adoption in pharmaceutical industry and related issues: A review. Journal of Clinical and Diagnostic Research. 2009;**3**:1942–1952

[3] Agarwal S, Mehrotra R. An overview of molecular docking. JSM Chemistry. 2016;**4**(2):1024

[4] Dnyandev KM, Babasaheb GV, Chandrashekhar KV, Chandrakant MA, Vasant OK. A review on molecular docking. International Research Journal of Pure and Applied Chemistry. 2021;**20**:60–68

[5] Raval K, Ganatra T. Basics, types and applications of molecular docking: A review. IP International Journal of Comprehensive and Advanced Pharmacology. 2022;**7**(1):12–16

[6] Náray-Szabó G. Protein Modelling. Cham, Switzerland: Springer International Publishing; 2014. 1–329 p

[7] Paggi JM, Pandit A, Dror RO. The art and science of molecular docking. Annual Review of Biochemistry. 2024;**93**:389–410

[8] Pinzi L, Rastelli G. Molecular docking: Shifting paradigms in drug discovery. International Journal of Molecular Sciences. 2019;**20**(18):4331

[9] Thakur G, Chaudhary S, Kumar Verma K. Drug repurposing and its various approaches. Int J Future Med Res. 2024;**6**(6):1–14

[10] Krishnamurthy N, Grimshaw AA, Axson SA, Choe SH, Miller JE. Drug repurposing: A systematic review on root causes, barriers and facilitators. BMC Health Services Research. 2022;**22**(1):970

[11] Roessler HI, Knoers NVAM, van Haelst MM, van Haaften G. Drug repurposing for rare diseases. Trends in Pharmacological Sciences. 2021;**42**(4):255–267

[12] Galindez G, Matschinske J, Rose TD, Sadegh S, Salgado-Albarrán M, Späth J, et al. Lessons from the COVID-19 pandemic for advancing computational drug repurposing strategies. Nature Computational Science. 2021;**1**:33–41

[13] Cousins HC, Nayar G, Altman RB. Computational approaches to drug repurposing: Methods, challenges, and opportunities. Annual Review of Biomedical Data Science. 2025;**7**:15–29

[14] Ballante F, Kooistra AJ, Kampen S, de Graaf C, Carlsson J. Structure-based virtual screening for ligands of G protein–coupled receptors: What can molecular docking do for you? Pharmacological Reviews. 2021;**73**(4):527–656

[15] Santana K, Do Nascimento LD, Lima E Lima A, Damasceno V, Nahum C, Braga RC, et al. Applications of virtual screening in bioprospecting: Facts, shifts, and perspectives to explore the chemo-structural diversity of natural products. Frontiers in Chemistry. 2021;**9**:1–23

[16] Zhang B, Li H, Yu K, Jin Z. Molecular docking-based computational platform for high-throughput virtual screening. CCF Transactions on High Performance Computing. 2022;**4**(1):63–74

[17] Fan J, Fu A, Zhang L. Progress in molecular docking. Quantitative Biology. 2019;7(2):83–89

[18] Feyza Maden S, Sezer S, Acuner SE. Fundamentals of Molecular Docking and Comparative Analysis of Protein-Small-Molecule Docking Approaches. In: Salih Istifli E, editor. Molecular Docking – Recent Advances. London, UK: IntechOpen; 2023. p. 1–24

[19] Dudhat K. Panoramic review on progress and development of molecular docking. Pharmaceutical Science and Technology. 2023;7(1):1–4

[20] Arya H, Coumar MS. Design of novel ROCK inhibitors using fragment-based de novo drug design approach. Journal of Molecular Modeling. 2020;**26**(9):249

[21] Alberts IL, Todorov NP, Dean PM. Receptor flexibility in de novo ligand design and docking. Journal of Medicinal Chemistry. 2005;**48**(21):6585–6596

[22] Brain CT, Chopra R, Kim S, Howard S, Sung MJ. The Discovery of Kisqali® (Ribociclib): A CDK4/6 Inhibitor for the Treatment of HR +/HER2- Advanced Breast Cancer. In: Wermuth CG, editor. Successful Drug Discovery: Volume 5. Hoboken, NJ, USA: Wiley; 2021. p. 273–289

[23] Rastelli G, Del Rio A, Degliesposti G, Sgobba M. Fast and accurate predictions of binding free energies using MM-PBSA and MM-GBSA. Journal of Computational Chemistry. 2010;**31**(4):797–810

[24] Defelipe LA, Arcon JP, Modenutti CP, Marti MA, Turjanski AG, Barril X. Solvents to fragments to drugs: MD applications in drug design. Molecules. 2018;**23**:3269

[25] Zhao Z, Bourne PE. Harnessing systematic protein–ligand interaction fingerprints for drug discovery. Drug Discovery Today. 2022;**27**(10):1–10

[26] Shoichet BK, Leach AR, Kuntz ID. Ligand solvation in molecular docking. Proteins: Structure, Function, and Genetics. 1999;**34**(1):4–16

[27] Huang SY, Zou X. Advances and challenges in Protein-ligand docking. International Journal of Molecular Sciences. 2010;**11**:3016–3034

[28] Ding QM, Huang YM, Yuan X. Molecular docking via quantum approximate optimization algorithm. Physical Review Applied. 2024;**213**:034036

[29] Li Y, Cui X, Xiong Z, Liu B, Wang BY, Shu R, et al. Quantum molecular docking with quantum-inspired algorithm. Journal of Chemical Theory and Computation. 2024;**20**:6687–6694

[30] Wang G, Zhu W. Molecular docking for drug discovery and development: A widely used approach but far from perfect. Future Medicinal Chemistry. 2016;**8**(14):1707–1710

[31] Huang N, Shoichet BK, Irwin JJ. Benchmarking sets for molecular docking. Journal of Medicinal Chemistry. 2006;**49**(23):6789–6801

[32] Lagarde N, Zagury JF, Montes M. Benchmarking data sets for the evaluation of virtual ligand screening methods: Review and perspectives. Journal of Chemical Information and Modeling. 2015;**55**:1297–1307

[33] Ding X, Chen X, Sullivan EE, Shay TF, Gradinaru V. Fast, accurate ranking of engineered proteins by target-binding propensity using structure modeling. Molecular Therapy. 2024;**32**(6):1687–1700

[34] Kazlauskas RJ, Bornscheuer UT. Finding better protein engineering strategies. Nature Chemical Biology. 2009;**5**:526–529

[35] Maldonado MR, Alnoch RC, de Almeida JM, Dos Santos LA, Andretta AT, Del Ropaín RPC, et al. Key mutation sites for improvement of the enantioselectivity of lipases through protein engineering. Biochemical Engineering Journal. 2021;**172**:108047. 108047

[36] Cao L, Coventry B, Goreshnik I, Huang B, Sheffler W, Park JS, et al. Design of protein-binding proteins from the target structure alone. Nature. 2022;**605**:551–560

[37] Patel J, Patel D, Meshram D. Artificial intelligence in pharma industry-a rising concept. Journal of Advanced Pharmacognosy. 2021; **1**(2):54–64

[38] Xu Z, Ren F, Wang P, Cao J, Tan C, Ma D, et al. A generative AI-discovered TNIK inhibitor for idiopathic pulmonary fibrosis: A randomized phase 2a trial. Nature Medicine. 2025;**31**:2602–2610

[39] Khamis MA, Gomaa W, Ahmed WF. Machine learning in computational docking. Artificial Intelligence in Medicine. 2015;**63**: 135–152

[40] García-Ortegón M, Seal S, Rasmussen C, Bender A, Bacallado S. Graph neural processes for molecules: An evaluation on docking scores and strategies to improve generalization. Journal of Cheminformatics. 2024;**16**:115

[41] Sánchez-Cruz N. Deep graph learning in molecular docking: Advances and opportunities. Artificial Intelligence in Life Sciences. 2023;**3**:100062

[42] Shen C, Ding J, Wang Z, Cao D, Ding X, Hou T. From machine learning to deep learning: Advances in scoring functions for protein–ligand docking. WIREs Computational Molecular Science. 2020;**10**:1429

[43] Crampon K, Giorkallos A, Deldossi M, Baud S, Steffenel LA. Machine-learning methods for ligand–protein molecular docking. Drug Discovery Today. 2022;**27**:151–164

[44] García-Ortegón M, Simm GNC, Tripp AJ, Hernández-Lobato JM, Bender A, Bacallado S. DOCKSTRING: Easy molecular docking yields better benchmarks for ligand design. Journal of Chemical Information and Modeling. 2021;**62**:3486–3502

[45] Gentile F, Yaacoub JC, Gleave J, Fernandez M, Ton AT, Ban F, et al. Artificial intelligence–enabled virtual screening of ultra-large chemical libraries with deep docking. Nature Protocols. 2022;**17**:672–697

[46] Tran-Nguyen VK, Jacquemard C, Rognan D. LIT-PCBA: An unbiased data set for machine learning and virtual

screening. Journal of Chemical Information and Modeling. 2020;**60**:4263–4273

[47] Wan Z, Sun X, Li Y, Chu T, Hao X, Cao Y, et al. Applications of artificial intelligence in drug repurposing. Advanced Science. 2025;**12**:2411325

[48] Zulhafiz NA, Teoh TC, Chin AV, Chang SW. Drug repurposing using artificial intelligence, molecular docking, and hybrid approaches: A comprehensive review in general diseases vs Alzheimer's disease. Computer Methods and Programs in Biomedicine. 2025;**261**:108604

[49] Choudhury C, Arul Murugan N, Priyakumar UD. Structure-based drug repurposing: Traditional and advanced AI/ML-aided methods. Drug Discovery Today. 2022;**27**(7):1847–1861

[50] Levin JM, Oprea TI, Davidovich S, Clozel T, Overington JP, Vanhaelen Q, et al. Artificial intelligence, drug repurposing and peer review. Nature Biotechnology. 2020;**38**:1127–1131

Chapter 2

Computational Chemistry Approaches to Molecular Docking

Bandenahalli Siddalingappa Krishna

Abstract

This chapter explores the diverse applications of molecular docking in computational chemistry, highlighting its central role in modern drug discovery and molecular research. Molecular docking predicts protein ligand interactions, providing insights into molecular recognition, binding affinity, and reaction mechanisms. By simulating how molecules fit and interact, docking aids in identifying potential drug candidates through virtual screening, significantly accelerating early-stage drug development. Beyond drug discovery, docking guides enzyme engineering by predicting modifications that enhance catalytic activity, enabling rational design of biocatalysts. It also supports studies of molecular recognition, helping researchers understand the structural and energetic factors underlying binding specificity and selectivity. By estimating binding energies, docking informs the prioritization of compounds for experimental testing, reducing time and costs. Recent advances have expanded docking's capabilities. Hybrid quantum mechanics/molecular mechanics (QM/MM) approaches improve the accuracy of binding predictions by accounting for electronic effects, while integration with molecular dynamics simulations captures protein flexibility and dynamic interactions. These enhancements allow more realistic modeling of biological systems and complex chemical reactions. Overall, molecular docking serves as a versatile tool across pharmaceutical, biochemical, and chemical sciences. Its ability to efficiently model interactions in silico accelerates research, supports rational design, and guides experimental efforts, making it indispensable in contemporary molecular research.

Keywords: molecular docking, protein–ligand interactions, drug design, enzyme engineering, virtual screening

1. Introduction

In recent decades, the rapid advancement of computational power, molecular modeling tools, and structural biology has fundamentally transformed the field of chemistry, giving rise to what we now refer to as **computational chemistry**. This discipline integrates principles of chemistry, physics, and computer science to model, simulate, and predict the behavior of molecular systems with remarkable precision. Among the most impactful techniques within this field is **molecular docking**, which enables scientists to predict how molecules – particularly small ligands – interact with biological macromolecules such as proteins, enzymes, DNA, or ribonucleic acid (RNA) [1, 2].

Molecular docking serves as a cornerstone methodology in structure-based drug design (SBDD) and has significantly influenced modern pharmaceutical research by allowing scientists to rapidly screen large compound libraries, evaluate binding affinities, and propose optimal drug candidates with specific molecular targets [3, 4]. It is especially valuable in the early stages of drug discovery, where time and cost constraints make experimental high-throughput screening impractical. With molecular docking, researchers can computationally predict and rank the binding poses and energies of thousands of compounds in a matter of hours.

In the context of **computational chemistry**, molecular docking plays a broader and more nuanced role beyond drug discovery. It is a **powerful analytical tool** for investigating **molecular recognition phenomena**, which are essential for understanding enzymatic activity, receptor signaling, metabolic pathways, and complex formation in biological systems. For example, by simulating how a substrate interacts with an enzyme's active site, docking can offer valuable insights into reaction mechanisms, catalytic efficiency, and potential allosteric effects [5, 6]. Such analyses contribute directly to the **rational design of enzyme variants** tailored for industrial applications, biocatalysts, or therapeutic use.

Another key area where molecular docking has proven indispensable is in the study of **protein–protein and protein–DNA interactions**. These interactions are fundamental to many cellular processes, including transcriptional regulation, immune response, and signal transduction. Docking simulations in these contexts allow researchers to identify interaction hotspots, predict binding interfaces, and even simulate the effects of mutations or structural changes on binding efficacy. When combined with tools such as **molecular dynamics (MD)** simulations and **free energy perturbation (FEP)** calculations, the predictive power of docking expands significantly, providing a more dynamic and realistic view of molecular behavior.

One of the reasons molecular docking has gained so much popularity in computational chemistry is its **balance of efficiency and accuracy**. While quantum chemical methods can offer high-precision results, they are often computationally expensive and not feasible for large biomolecular systems. Docking, in contrast, provides an efficient alternative that can be scaled to handle complex biological molecules and diverse chemical libraries, especially when used in tandem with machine learning (ML) algorithms or cloud-based screening platforms [7, 8].

Recent developments in the field have led to the integration of **hybrid quantum mechanics/molecular mechanics (QM/MM) methods**, which allow researchers to apply quantum calculations to the reactive site of interest while modeling the rest of the system with classical force fields [9]. This combination offers a deeper understanding of the **electronic environment** and **reaction pathways** involved in ligand binding or enzymatic catalysis. Additionally, **AI-driven approaches** and **deep learning models** are beginning to reshape the way docking is performed, offering enhanced scoring functions, pose prediction, and the ability to learn from large datasets of known interactions.

As computational tools become more accessible and user-friendly, molecular docking is being increasingly adopted by researchers outside traditional chemistry departments – such as in **biomedical engineering**, **pharmaceutical sciences**, **materials science**, and even **agricultural biotechnology**. This widespread applicability underscores the versatility of docking as a scientific tool and highlights the need for a deeper understanding of its capabilities and limitations.

This chapter explores the applications of molecular docking, specifically within the field of computational chemistry, covering its use in drug discovery, enzyme engineering, protein interaction studies, and reaction mechanism analysis. It also discusses the tools and software commonly used, the theoretical principles behind docking, and the emerging trends that are shaping the future of in silico molecular modeling.

By the end of this chapter, readers will gain a solid understanding of how molecular docking is applied in computational chemistry to address real-world scientific challenges, reduce experimental workloads, and accelerate the pace of innovation in both academic and industrial research settings [10, 11].

2. Fundamentals of molecular docking

2.1 What is molecular docking?

Molecular docking is a computational technique used to predict the preferred orientation and binding affinity of one molecule (typically a small ligand) when it interacts with a second molecule (usually a larger biological macromolecule such as a protein or DNA). The goal is to determine how two molecules fit together in a **receptor–ligand complex**, much like how a key fits into a lock.

The process simulates the interaction between the molecules to find the most stable configuration – known as the **binding pose** – which often correlates with **biological activity** or **inhibitory potential**. The docking process evaluates multiple orientations and conformations of the ligand in the binding site and scores them based on factors like hydrogen bonding, hydrophobic interactions, van der Waals forces, and electrostatics [12, 13].

Molecular docking is widely used in:

- **Structure-based drug design**: to identify promising drug candidates
- **Enzyme-substrate interaction studies**: to model catalysis and inhibition
- **Biomolecular interaction mapping**: for protein–protein, protein–DNA, and protein–RNA complexes

It is considered a fast and cost-effective alternative to experimental methods, especially in the early stages of drug discovery and molecular analysis.

A dagram (**Figure 1**) illustrating the molecular docking workflow in computational drug discovery, showing steps such as protein and ligand preparation, docking simulation, scoring and ranking of binding poses and analysis of praotein-lingand interactions.

Protein of interest: at the top, the figure shows a **protein structure** (ribbon model with helices, sheets, and coils). This protein is the **target** for docking, usually an enzyme, receptor, or other biomolecule involved in a disease pathway. The protein has an **active/binding site** (cavity or pocket) where potential ligands can interact as shown in the **Figure 1**.

Chemical database: the middle section represents a library of chemical compounds (ligands). This database contains thousands to millions of small molecules with diverse chemical structures. The goal is to screen them

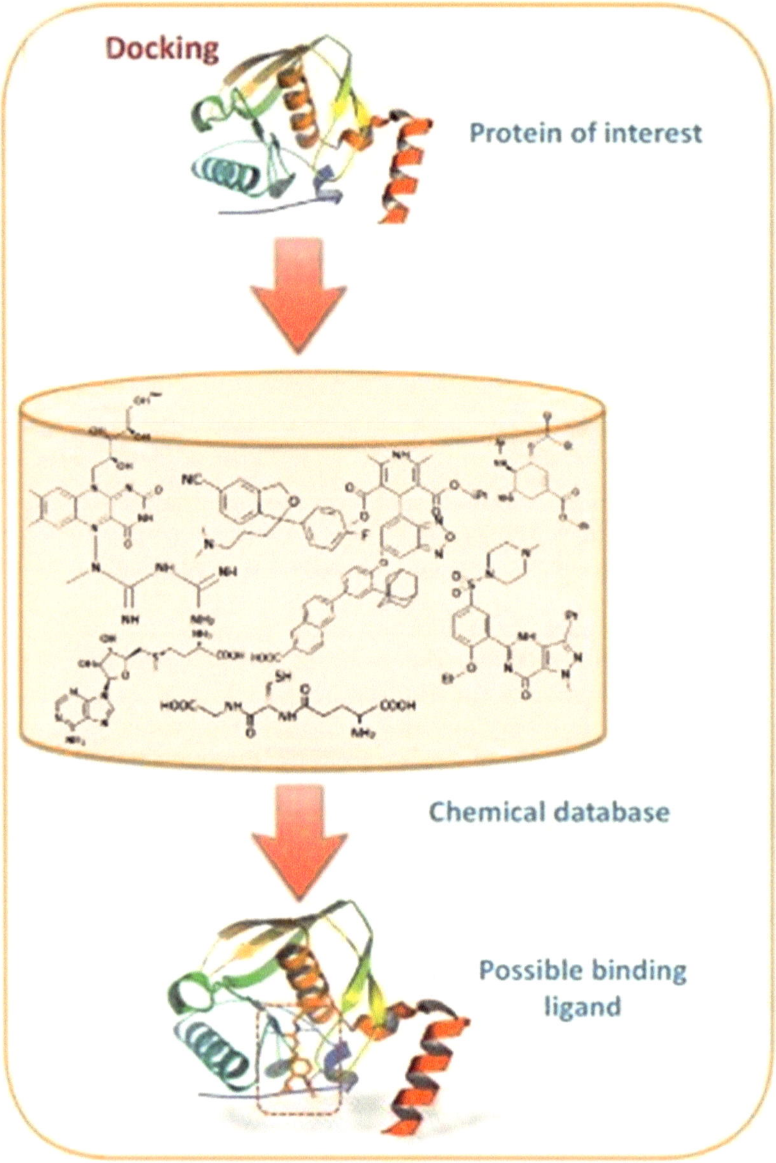

Figure 1.
Molecular docking workflow used in computational drug discovery to study interactions between proteins and ligands.

computationally to determine which ones might fit and bind well into the protein's active site.

Docking process: docking is a simulation technique where each chemical compound is virtually "fitted" into the protein's binding site. It predicts binding orientation (how the ligand fits into the site), binding affinity (strength of interaction, usually expressed as docking score or binding energy), and key molecular interactions (hydrogen bonds, hydrophobic contacts, electrostatic interactions, π-π stacking, van der Waals forces, etc.).

Possible binding ligand: at the bottom, the figure shows the **protein**-ligand complex after docking. The ligand is predicted to **bind stably** within the protein's active site. Such a ligand is a **potential drug candidate**, as it can modulate the protein's activity (inhibit or activate). This **Figure 1** illustrates the workflow of SBDD. The process begins with selecting a protein target of biomedical relevance, followed by virtual screening (VS) of a chemical database through molecular docking. From this, potential ligands with favorable binding interactions are identified for further evaluation.

The **Figure 1** explains how molecular docking helps identify new drug candidates by simulating and predicting how different chemical compounds might interact with a target protein at the molecular level. It reduces time and cost compared to experimental screening.

2.2 Types of molecular docking

Molecular docking can be categorized based on the flexibility of the molecules involved:

2.2.1 Rigid docking

In rigid docking, both the ligand and the receptor are treated as inflexible structures. This method assumes that the molecular conformations are fixed, and only spatial positioning is adjusted. Although computationally efficient, it often lacks biological accuracy because real molecules are flexible and can adapt upon binding.

2.2.2 Flexible docking

Flexible docking allows conformational changes in either the ligand, the receptor, or both. This approach is more realistic as it simulates how molecules behave dynamically in a biological system. There are different levels of flexibility:

- **Ligand-flexible docking**: only the ligand's conformation changes

- **Receptor-flexible docking**: allows side-chain or backbone movements in the protein

- **Induced fit docking**: both molecules adapt to achieve optimal binding

Flexible docking is more computationally demanding but provides a better approximation of real molecular interactions.

A diagram (**Figure 2**) showing types of molecular docking studies categorized by receptor and ligand flexibility, including rigid–rigid docking, flexible–ligand docking, flexible–receptor docking, and fully flexible docking approaches.

Rigid docking: in rigid docking, both the receptor and the ligand are treated as fixed, inflexible structures. The assumption is that neither molecule undergoes conformational changes during the binding process. This method is computationally efficient and requires fewer resources, making it suitable for high-throughput screening or initial binding predictions. However, because biological molecules are inherently flexible, rigid docking often lacks accuracy in representing true binding interactions.

Flexible–rigid docking: flexible–rigid docking is a more refined approach, where the receptor is kept rigid while the ligand is allowed to adopt different conformations. This

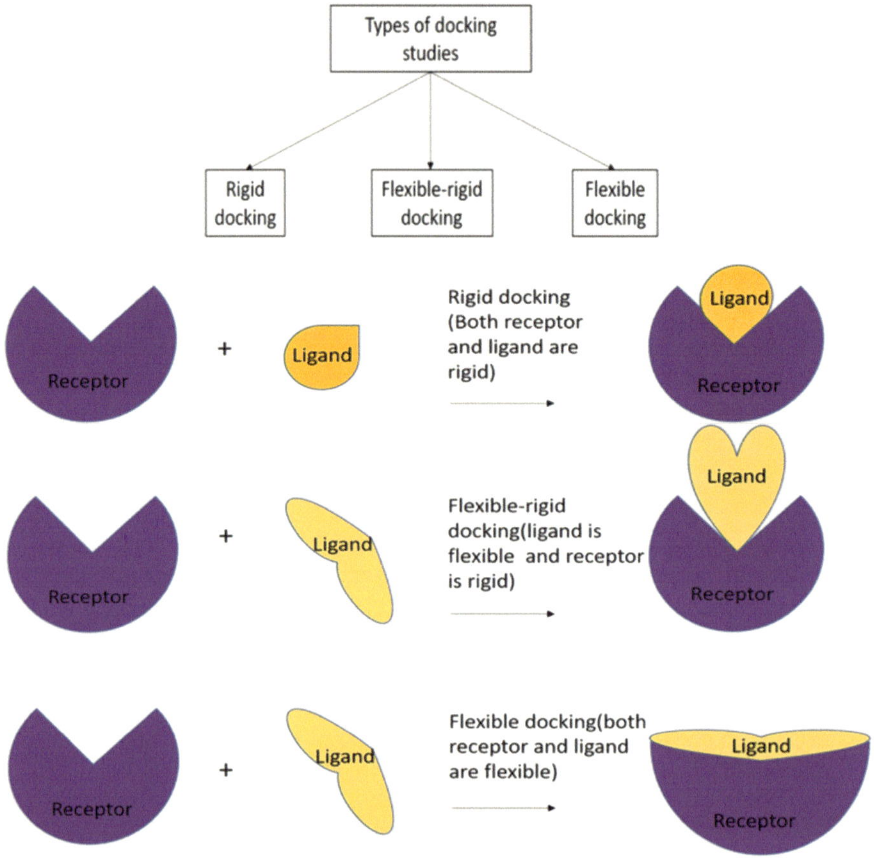

Figure 2.
Types of molecular docking studies based on the flexibility of the receptor and ligand.

flexibility enables the ligand to adjust its shape to fit into the binding pocket of the receptor. As a result, this method provides a more realistic prediction of binding interactions compared to rigid docking. It is one of the most commonly used approaches in SBDD because it balances computational efficiency with predictive accuracy.

Flexible docking: flexible docking is the most advanced and realistic method, as it allows both the receptor and the ligand to undergo conformational changes during the docking process. This approach as shown in **Figure 2** accounts for the dynamic nature of proteins and small molecules, thereby providing highly accurate predictions of binding affinity and orientation. However, the increased accuracy comes at the cost of high computational demand and longer processing times. Flexible docking is particularly valuable for complex systems where receptor flexibility plays a key role in ligand recognition and binding.

The choice of docking method depends on the balance between speed, computational resources, and accuracy required for a given study. Rigid docking is fast but less accurate, flexible–rigid docking provides a practical compromise, and flexible docking offers the most accurate yet computationally expensive predictions. Together, these approaches form the foundation of SBDD and play a critical role in modern drug discovery research.

2.3 Scoring functions in docking

Scoring functions are mathematical models used to **predict the strength and stability of the ligand–receptor interaction**. After generating multiple binding poses, the docking algorithm uses these functions to rank and select the best pose based on predicted binding affinity [4, 9, 12, 13].

Common types of scoring functions:

1. **Force-field-based scoring** calculates interaction energies based on physical principles (e.g., van der Waals, electrostatics). *Example tools: AutoDock, CHARMM.*

2. **Empirical scoring** combines physical terms with regression models derived from experimental data. *Example: ChemScore, GlideScore.*

3. **Knowledge-BASED SCORING** uses statistical data from known protein–ligand complexes to estimate interaction probabilities. *Example: DrugScore*

4. **Consensus scoring** combines results from multiple scoring functions to improve accuracy.

While scoring functions are essential, they are often approximate and may fail to capture all energetic and entropic contributions. Therefore, **postdocking analyses**, such as MD or free energy calculations, are often used for refinement [14].

3. Docking algorithms and software tools

The success of molecular docking relies heavily on the efficiency of the algorithms used to explore ligand–receptor interactions and identify optimal binding poses. Docking algorithms are designed to simulate the binding process by sampling many possible orientations, positions, and conformations of the ligand within the receptor's active site. These are then evaluated using scoring functions to determine the best fit.

3.1 Docking algorithms

Docking algorithms typically follow a two-step process:

- **Search algorithm**: explores different binding modes between the ligand and the receptor.

- **Scoring function**: evaluates and ranks the binding modes based on the predicted binding affinity.

3.1.1 Common types of search algorithms

- **Systematic search:** explores all possible conformations and orientations of the ligand exhaustively. It provides accurate results but is computationally intensive.

- **Stochastic methods:** use random or probabilistic sampling techniques (e.g., Monte Carlo simulations, genetic algorithms). These are widely used for their balance between speed and accuracy.

- **Molecular dynamics-based search:** uses time-dependent simulations to model ligand movement and flexibility. It is often coupled with other techniques to refine docking poses.

- **Fragment-based docking:** ligands are broken into fragments and docked sequentially. This approach is useful for building novel ligands or optimizing binding affinity.

3.2 Common molecular docking software tools

Common molecular docking software tools—such as AutoDock, AutoDock Vina, DOCK, Glide, and GOLD (as shown in the **Table 1**)—are widely used in drug discovery and virtual screening to predict how small molecules bind to protein targets. These programs differ in speed, accuracy, flexibility handling, and ease of use. Open-source tools like AutoDock and Vina are popular in academia because they are free, support ligand and limited receptor flexibility, and are suitable for large-scale screening. Commercial tools such as Glide and GOLD often offer higher accuracy and more sophisticated scoring functions, making them useful for detailed binding studies. The choice of software depends on research goals, computational resources, and the level of expertise required. Below is a list of widely used molecular docking programs, each with its own features, algorithms, and applications:

Each of these tools balances **speed, accuracy, and usability** differently, and the choice of software depends on the research goals, computational resources, and level of expertise.

Software tool	Key features	Usage/application
AutoDock and AutoDock Vina	Open-source; supports ligand and receptor flexibility; widely used in academia	Drug discovery, virtual screening
Glide (Schrödinger)	Commercial: high-precision docking with advanced scoring	Lead optimization, pharmaceutical R&D
GOLD (CCDC)	Genetic algorithm-based docking allows for full ligand flexibility	Protein–ligand docking, virtual screening
DOCK (UCSF)	One of the earliest docking programs: flexible ligand docking	Structure-based drug design
SwissDock	Web-based tool using EADock algorithm	Quick and accessible docking for researchers
FlexX	Fast incremental construction algorithm	Known for handling ligand flexibility efficiently

Table 1.
Strengths and limitations of each software.

4. Applications in drug discovery

Molecular docking has become an indispensable tool in the field of **drug discovery and development**, particularly within the **SBDD** paradigm. By predicting how potential drug molecules bind to a specific biological target, docking accelerates the identification and optimization of therapeutic candidates – dramatically reducing time, cost, and the need for extensive wet-lab experimentation.

4.1 Virtual screening of drug libraries

VS is one of the most impactful applications of molecular docking. It involves screening thousands to millions of compounds in silico to identify those most likely to bind effectively with a target protein. Docking is used to:

- Rank compounds based on predicted binding affinity
- Select top hits for synthesis or experimental testing
- Eliminate nonbinders early in the pipeline

There are two major types of VS:

- **Structure-based VS**: uses the 3D structure of the target (via X-ray crystallography, NMR, or homology modeling)
- **Ligand-based VS**: relies on known active compounds to predict new candidates when the target structure is unknown

This method is widely used in pharmaceutical industries to fast-track hit identification and reduce false positives in drug discovery pipelines [15, 16, 17].

A workflow diagram (**Figure 3**) illustrating the virtual screening and docking process, including compound library preparation, filtering and selection, docking simulations, scoring and ranking of hits, and identification of promising lead molecules. The process of structure-based drug discovery often begins with the construction of a virtual compound library. In this step, a collection of small molecules with diverse chemical structures is prepared, either from publicly available chemical databases or through computational design. This library serves as the source of potential ligands that can interact with the biological target.

The next step is molecular docking, where the compounds in the virtual library are computationally fitted into the active site of the target protein. Docking simulations predict the binding orientation, interactions, and binding affinity of each ligand with the receptor. This step helps identify how well a compound can occupy the binding pocket and establish stabilizing interactions, such as hydrogen bonds, hydrophobic interactions, and electrostatic forces as shown in **Figure 3**.

Following docking, the results undergo VS, in which compounds are ranked based on their predicted binding scores and interaction profiles. This filtering process allows researchers to shortlist the most promising ligands with favorable binding characteristics for further experimental validation. By integrating these steps, the workflow significantly reduces time and cost in the early stages of drug discovery while enhancing the likelihood of identifying effective drug candidates.

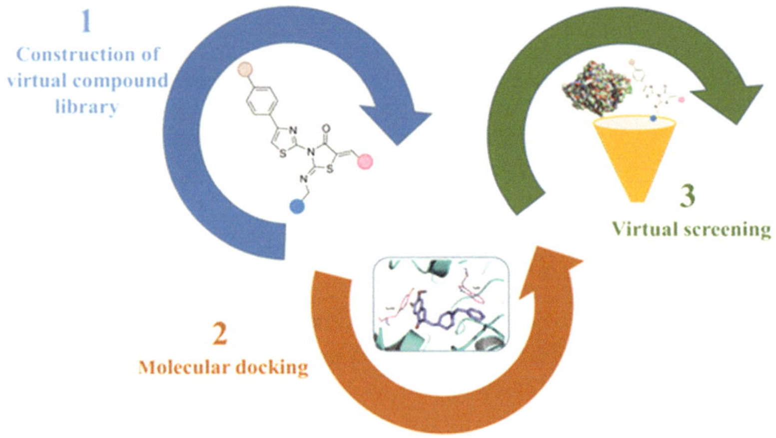

Figure 3.
Workflow of virtual screening and docking.

4.2 Lead optimization

Once a potential drug candidate (or **lead compound**) is identified, docking helps in **optimizing its structure** to improve potency, selectivity, and pharmacokinetic properties. Iterative docking simulations are performed on modified compounds to:

- Improve binding strength (affinity)
- Reduce off-target effects
- Enhance specificity toward a receptor subtype

Tools like Glide and GOLD are particularly useful for this purpose, as they offer high-precision scoring and flexibility in ligand docking.

4.3 Predicting binding modes and interactions

Docking helps visualize how a drug binds to its target, revealing:

- Key **binding site residues**
- Types of interactions (hydrogen bonds, hydrophobic contacts, π–π stacking, etc.)
- **Orientation and conformation** of the ligand within the active site

This information is critical in **rational drug design**, where specific features of a compound are engineered to exploit the structure of the binding pocket [6, 15].

A graphic showing, in **Figure 4**, how molecular dynamics (MD) simulations validate docking results by tracking the ligand–protein complex over time and using RMSD plots to confirm stable binding interactions. After docking, it is

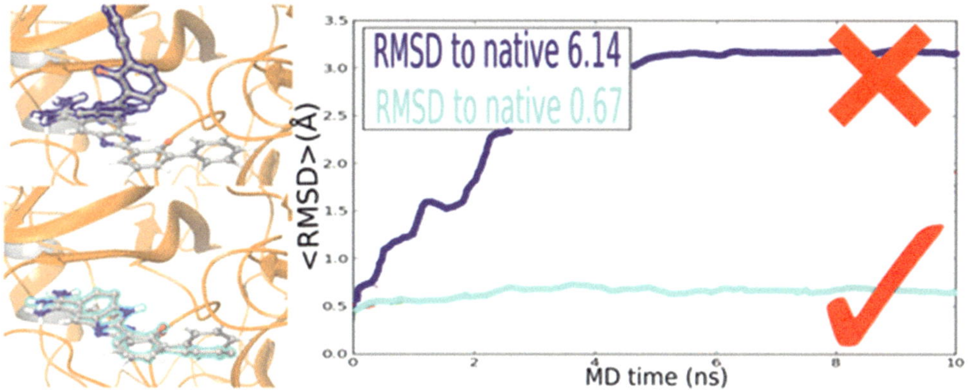

Figure 4.
MD simulations validate docking by identifying stable ligand–protein interactions through RMSD.

essential to evaluate the stability of the predicted ligand–protein complexes under dynamic conditions. MD simulations provide valuable insights into how a ligand behaves in the binding pocket over time. One of the most commonly used parameters in this analysis is the root mean square deviation (RMSD), which measures the deviation of atomic positions compared to the native (reference) structure as shown in **Figure 4**.

In the **Figure 4**, two docking poses are compared based on their RMSD values. The blue trajectory shows a high RMSD value of 6.14 Å, indicating poor stability and significant deviation from the native binding mode. Conversely, the cyan trajectory shows a low RMSD value of 0.67 Å, which suggests strong stability and close agreement with the native structure. The structural snapshots on the left further illustrate how the correct pose remains stably bound within the active site, while the incorrect pose drifts away.

Thus, RMSD analysis during MD simulations is a critical step in distinguishing reliable binding modes from false positives, thereby improving the accuracy of structure-based drug discovery.

4.3.1 Repurposing existing drugs

Docking is also used in **drug repurposing**, where approved or shelved drugs are computationally screened against new targets. This approach:

- Reduces development time and cost
- Explores new therapeutic indications for existing compounds
- Was especially useful during **emergency health crises** (e.g., COVID-19 drug-screening efforts)

4.4 Case studies and real-world impact

Several successful drugs were first identified or optimized using molecular docking methods:

- **HIV protease inhibitors** (e.g., saquinavir) were developed using structure-guided docking
- **Imatinib** (Gleevec), used in chronic myeloid leukemia, benefited from docking in kinase binding analysis
- **COVID-19 antiviral candidates** were rapidly screened using AutoDock and other tools

These real-world examples highlight how molecular docking transforms drug discovery from a trial-and-error process into a **targeted, hypothesis-driven** workflow.

5. Applications in enzyme engineering

Beyond drug discovery, molecular docking plays a vital role in **enzyme engineering**, where the goal is to understand, improve, or redesign enzyme activity for specific applications in medicine, biotechnology, and industry. Docking allows researchers to explore how substrates, inhibitors, and cofactors interact with enzyme active sites, offering insights into both natural function and opportunities for rational design.

5.1 Understanding enzyme–substrate interactions

Molecular docking is frequently used to study how enzymes recognize and bind their natural substrates. By simulating these interactions, researchers can:

- Identify **key active site residues**
- Visualize the **orientation** of the substrate within the catalytic pocket
- Understand **binding specificity** and **reaction mechanism pathways**

This knowledge forms the foundation for modifying enzyme selectivity or improving substrate turnover in industrial or therapeutic enzymes.

A diagram (**Figure 5**) illustrating the formation of the enzyme–substrate complex, showing a substrate molecule fitting into the enzyme's active site to form a temporary bound complex before catalysis occurs. Enzymes are biological catalysts that accelerate chemical reactions by binding to specific substrates. Each enzyme contains an active site, a region with a unique shape complementary to the substrate. The substrate fits into the active site much like a key fits into a lock, ensuring specificity of the reaction.

When the substrate binds to the enzyme's active site, they form an enzyme–substrate complex as shown in **Figure 5**. This complex stabilizes the transition state and lowers the activation energy required for the reaction. As a result, the reaction proceeds more efficiently, converting the substrate into the product. Once the reaction is complete, the enzyme releases the product and is free to catalyze another reaction cycle.

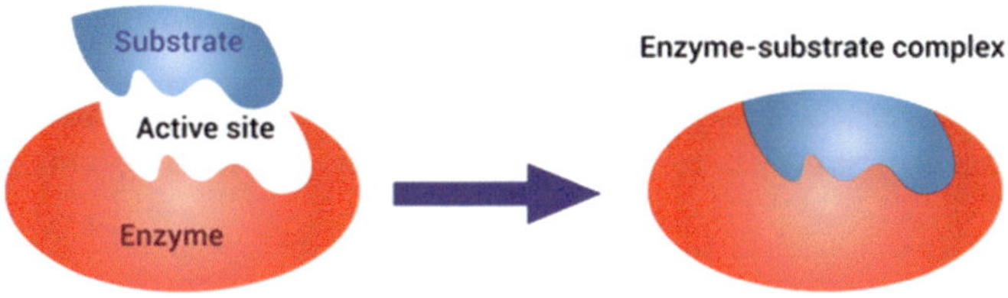

Figure 5.
Formation of the enzyme–substrate complex.

This **Figure 5** shows the substrate binding to the enzyme's active site and the subsequent formation of the enzyme–substrate complex, highlighting the basis of enzyme catalysis and specificity.

5.2 Rational enzyme design

With molecular docking, scientists can rationally design enzymes to:

- Accept nonnatural substrates
- Improve catalytic efficiency
- Alter regioselectivity or stereoselectivity
- Enhance stability under extreme conditions (e.g., temperature, pH)

By docking modified substrates or potential transition states into mutant enzyme models, researchers can predict how amino acid changes affect binding and catalysis – before doing any experimental work.

Example: engineering lipases or esterases to better accept bulky ester substrates for use in biofuel production.

5.3 Enzyme inhibition and allosteric modulation

Molecular docking helps in designing or analyzing enzyme inhibitors – whether competitive, noncompetitive, or allosteric. It can:

- Identify potential binding pockets outside the active site
- Predict conformational changes upon binding
- Guide the design of small-molecule regulators for therapeutic enzymes

For example, docking has been used to identify allosteric inhibitors of kinases and proteases, offering a strategy to modulate enzyme function without directly competing with substrates.

5.4 Integration with mutagenesis and experimental validation

In silico docking studies are often paired with **site-directed mutagenesis** and **enzyme kinetics experiments** to validate and refine predictions. This synergy allows for:

- Prioritizing mutations for laboratory testing
- Reducing the number of experimental iterations
- Gaining structural explanations for observed functional changes

5.5 Case study: Docking in industrial biocatalysts

A notable application is the use of docking in the development of **enzymes for industrial catalysis**, such as:

- **Proteases** used in detergents
- **Amylases** for starch processing
- **Dehydrogenases** for stereoselective synthesis

Molecular docking helps identify which enzyme variants bind better to synthetic or chiral substrates, thus accelerating the path from laboratory design to industrial application.

6. Modeling protein–protein and protein–DNA interactions

Molecular docking extends beyond small molecule–protein interactions to include **macromolecular complexes**, such as protein–protein and protein–DNA interactions. These complexes are fundamental to numerous biological processes, including signaling, transcription, and immune responses. Computational docking methods have evolved to tackle the unique challenges posed by these larger, flexible systems [18, 19].

A diagram (**Figure 6**) showing the central dogma of molecular biology, illustrating how DNA is transcribed into RNA and how RNA is then translated into protein, with arrows indicating the directional flow of genetic information. The genetic material of all living organisms is stored in the form of DNA. DNA has the unique ability to undergo replication, ensuring that genetic information is faithfully passed on to daughter cells during cell division.

Through the process of transcription, a segment of DNA is used as a template to synthesize RNA. RNA acts as an intermediate messenger that carries the genetic code from the DNA in the nucleus to the ribosomes in the cytoplasm. Interestingly, some viruses utilize reverse transcription, where RNA serves as the template to synthesize complementary DNA (cDNA), allowing them to integrate into host genomes.

The RNA sequence is then decoded through the process of translation, where ribosomes read the nucleotide sequence and assemble the corresponding sequence of

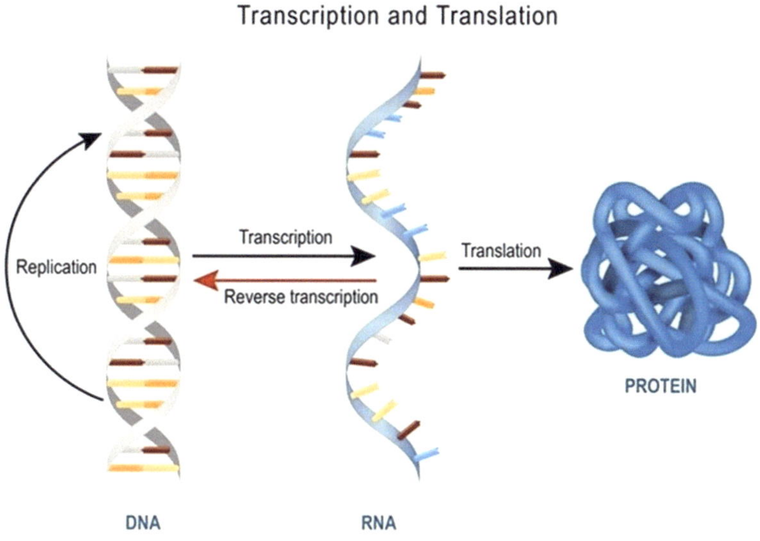

Figure 6.
The central dogma: transcription and translation.

amino acids as shown in **Figure 6**. This leads to the synthesis of proteins, which are essential biomolecules that perform structural, catalytic, regulatory, and transport functions within the cell.

This **Figure 6** summarizes the fundamental principle of molecular biology: DNA makes RNA, and RNA makes protein, thereby linking genetic information to functional molecules.

6.1 Importance of protein–protein and protein–DNA docking

- **Protein–protein interactions (PPIs):** regulate key cellular processes, such as signal transduction, metabolic pathways, and immune recognition.

- **Protein–DNA interactions:** control gene expression, replication, and repair mechanisms.

Understanding these interactions at the molecular level is crucial for designing **therapeutic inhibitors**, synthetic biology applications, and deciphering biological networks.

This image illustrates the workflow of protein–DNA/RNA docking interactions, which is a computational approach to predict how a protein binds to nucleic acids. Let's break it down step-by-step.

A workflow diagram illustrating the steps in protein–DNA/RNA docking interactions, including structure preparation of the protein and nucleic acid, selection of docking software, docking simulation, scoring and ranking of poses, and final analysis of predicted binding interactions.

Input molecules: protein (shown as a ribbon diagram in red) and DNA or RNA (shown as the helical structure). These two are the starting molecules whose interaction we want to model.

System preparation: before docking, the protein and DNA/RNA structures are cleaned, formatted, and pre-processed. This may include removing water molecules, adding missing atoms, assigning charges, and defining flexible regions.

Docking calculation: the prepared protein and nucleic acid are virtually positioned in various orientations and conformations using computational algorithms. The goal is to simulate all possible ways they could interact in three-dimensional space.

Scoring: each predicted binding pose is evaluated using a scoring function, which estimates the binding affinity or stability of the complex. The scoring considers hydrogen bonds, hydrophobic contacts, electrostatic interactions, and steric fit.

Complexed structure: the best-scoring docked conformation is selected as the predicted complex. This is the computationally generated model of how the protein interacts with DNA/RNA as shown in **Figure 7**.

Interaction types in such complexes

When proteins bind to DNA/RNA, common interactions include:

- **Hydrogen bonding** (between amino acid side chains and nucleotide bases or the phosphate backbone)

- **Electrostatic interactions** (positive residues like Lys and Arg with negatively charged phosphate groups)

Van der Waals contacts:

- **Base stacking** (aromatic amino acids interacting with nucleobases)

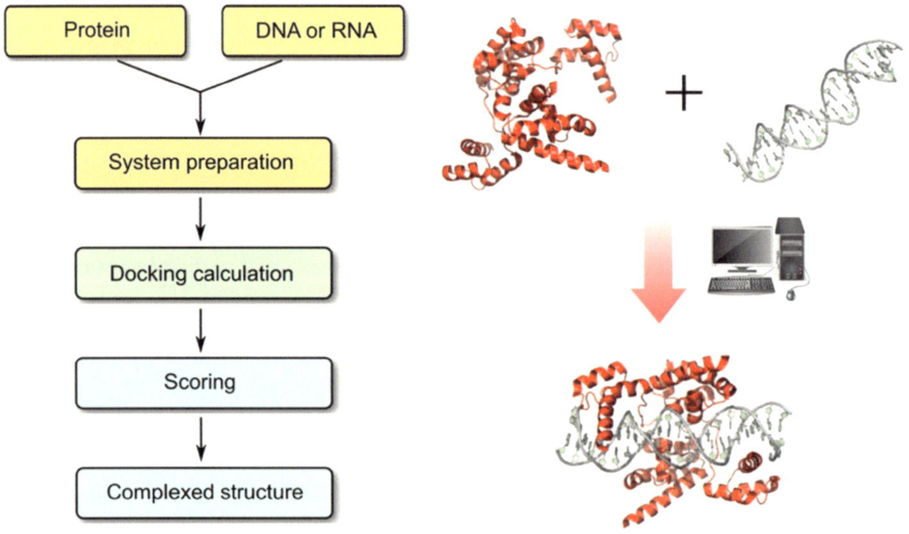

Figure 7.
Workflow of protein–DNA/RNA docking interactions.

6.2 Challenges in macromolecular docking

- Larger interfaces with extensive contact surfaces
- Significant conformational flexibility and induced fit upon binding
- Difficulties in accurately scoring and predicting binding affinity
- Presence of water molecules and cofactors affecting interactions

These complexities require specialized docking approaches and often integrate experimental data, such as mutagenesis or NMR restraints.

6.3 Docking approaches for protein–protein and protein–DNA complexes

- **Rigid-body docking:** assumes minimal conformational changes and uses shape complementarity to predict complex formation. Useful as a starting point.
- **Flexible docking:** incorporates side-chain or backbone flexibility using MD or ensemble docking techniques to capture induced fit.
- **Template-based docking:** uses known structures of similar complexes as models; helpful when homologous complexes are available.

6.4 Popular tools and algorithms

Several specialized software tools have been developed for macromolecular docking, including:

- **High Ambiguity Driven DOCKing (HADDOCK):** incorporates experimental data and flexible docking
- **ClusPro:** automated rigid-body protein–protein docking server
- **ZDOCK:** uses fast Fourier transform (FFT)-based rigid docking
- **Rosetta dock:** allows flexible docking with energy-based scoring
- **3D-DART:** specialized for protein–DNA docking

6.5 Applications

- Predicting signaling complexes in pathways like MAPK and apoptosis
- Designing inhibitors that block protein–protein interfaces in diseases such as cancer
- Understanding transcription factor binding to DNA for gene regulation studies

- Engineering synthetic transcription factors or DNA-binding proteins for biotechnology

7. Integration with molecular dynamics and QM/MM

While molecular docking provides valuable snapshots of ligand–receptor interactions, it often treats molecules as relatively rigid entities and relies on static scoring functions. To overcome these limitations and gain a deeper understanding of molecular recognition and binding energetics, docking is increasingly combined with MD simulations and QM/MM methods.

A diagram (**Figure 8**) showing the applications of QM/MM methods in biomolecular and nanomaterial studies, including enzyme reaction mechanisms, drug–target interactions, photochemical processes, catalytic activity in nanomaterials, and electronic property analysis, with the quantum region highlighted and the surrounding molecular environment treated by classical mechanics. The QM/MM approach is a powerful computational technique that combines the accuracy of QM with the efficiency of MM. It is particularly useful for investigating large and complex systems where a fully quantum treatment would be computationally prohibitive. In this method, the chemically active region (such as a metal center, ligand, or catalytic site) is modeled using QM, while the surrounding environment (protein, solvent, or bulk structure) is treated with MM.

Figure 8 shows diverse applications of QM/MM simulations:

- **Thiolated-protected Aun_nn:** QM/MM is used to study gold nanoclusters stabilized by thiol ligands, providing insights into their stability and optical properties.

- **Agn_nn + DNA:** quantum-level interactions between silver clusters and DNA are modeled to understand binding, structural stability, and electronic properties.

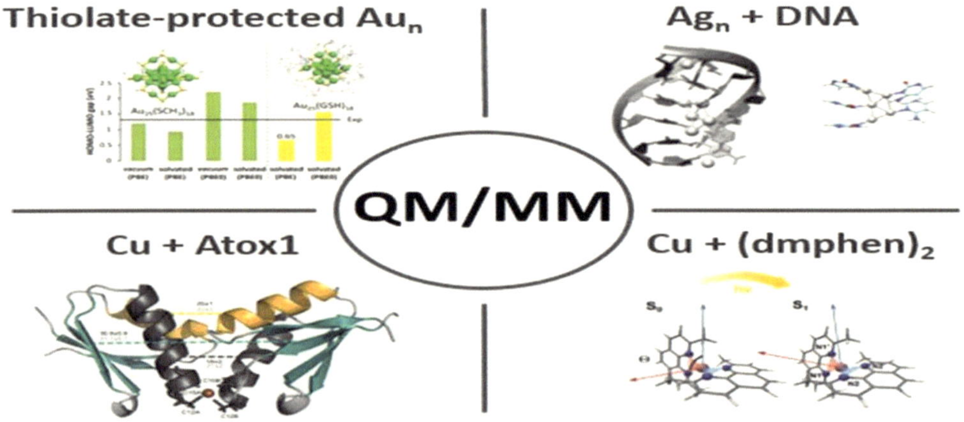

Figure 8.
Applications of QM/MM in biomolecular and nanomaterial studies.

- **Cu+Atox1:** the interaction of copper ions with the metallochaperone protein Atox1 is studied to reveal the structural and electronic basis of metal transport in biological systems.
- **Cu+ (dmphen)2_22:** photophysical properties of copper complexes, such as electronic transitions upon light absorption, are analyzed using QM/MM to understand their excited-state behavior.

By integrating quantum precision with molecular-scale modeling, QM/MM provides a balanced approach to studying metalloproteins, metal–ligand complexes, and nanomaterial–biomolecule interactions.

7.1 Molecular dynamics (MD) simulations

MD simulations model the time-dependent behavior of molecules by solving Newton's equations of motion, allowing researchers to:

- Explore the **conformational flexibility** of both the ligand and the receptor
- Observe **binding and unbinding events** dynamically
- Refine docking poses by allowing the relaxation of complexes in a simulated environment
- Calculate **binding free energies** through techniques like MM-PBSA and MM-GBSA

By integrating docking with MD, scientists can better predict realistic binding modes and improve the reliability of docking outcomes.

7.2 Quantum mechanics/molecular mechanics (QM/MM) methods

QM/MM combines the accuracy of quantum mechanical calculations for the reactive site with the efficiency of MM for the rest of the system. This hybrid approach is particularly useful when:

- Studying enzyme catalysis and reaction mechanisms
- Investigating electronic effects in ligand binding
- Modeling bond formation/breaking events that classical docking cannot capture

QM/MM can complement docking by providing detailed insight into transition states and energy barriers, which are critical for enzyme engineering and drug design.

7.3 Workflow integration

A common workflow includes:

- Initial docking to predict binding poses
- MD simulations to relax and explore the dynamics of the docked complex
- QM/MM calculations on the active site or ligand-binding region to analyze electronic structure and reaction pathways

This multiscale modeling approach enhances predictive power and enables comprehensive mechanistic studies.

7.4 Advantages and challenges

Advantages:

- Accounts for molecular flexibility and solvent effects
- Provides detailed energetic and mechanistic information
- Improves the accuracy of binding affinity predictions

Challenges:

- High computational cost, especially for QM/MM
- Requires expertise to set up and interpret simulations
- Integration complexity and large data handling

8. Challenges and limitations of molecular docking

Despite its significant contributions to computational chemistry and drug discovery, molecular docking faces several challenges and limitations that impact its accuracy and **applicability.**

8.1 Protein and ligand flexibility

One of the main challenges in docking is accurately accounting for the **flexibility** of both proteins and ligands. Most docking programs simplify the problem by treating the receptor as rigid or only partially flexible. However, proteins often undergo significant conformational changes upon ligand binding (induced fit), which rigid docking methods cannot capture effectively [19, 20]. Similarly, ligands with many rotatable bonds create an enormous conformational search space, complicating accurate predictions.

8.2 Scoring function limitations

Current scoring functions are approximate and often struggle to reliably predict absolute binding affinities or rank ligands correctly. They may:

- Overlook important entropic contributions
- Inaccurately model solvation and desolvation effects
- Fail to capture specific interactions, such as metal coordination or water-mediated hydrogen bonds

This can lead to false positives or negatives, affecting VS outcomes.

8.2.1 Treatment of solvent effects

Water molecules play a critical role in molecular recognition by mediating hydrogen bonds and affecting binding thermodynamics [21]. Most docking methods treat solvent implicitly or neglect it altogether, which limits the accuracy of predicted binding modes and energies.

8.2.2 Computational cost versus accuracy trade-off

More accurate methods that include receptor flexibility, explicit solvent, and advanced scoring functions often require significant computational resources. Balancing speed and accuracy remains a challenge, especially when screening large compound libraries.

8.2.3 Quality of structural data

Docking accuracy heavily depends on the quality of the receptor structure. Errors or missing regions in crystal structures, or uncertainties in homology models, can lead to incorrect binding predictions.

8.2.4 Lack of standardization

There is no universally accepted docking protocol or scoring function. Results can vary between software packages, making it essential to validate docking predictions with experimental data or complementary computational methods.

Despite these challenges, continuous advances in algorithms, ML, and hybrid approaches are improving docking reliability and expanding its applications.

9. Future perspectives and advances

Molecular docking continues to evolve rapidly, driven by advances in computational power, algorithms, and integration with experimental data. The future of docking promises to address current limitations and expand its role in biomedical and chemical research.

9.1 Artificial intelligence and machine learning integration

AI and ML are transforming docking by:

- Improving scoring functions using large datasets and pattern recognition
- Enhancing pose prediction accuracy through deep learning models
- Automating hit identification and lead optimization workflows

These approaches enable more reliable predictions with less human intervention, accelerating drug discovery.

9.2 Enhanced treatment of flexibility

Next-generation docking methods increasingly incorporate:

- Full protein and ligand flexibility through ensemble docking or MD-guided docking
- Modeling of induced fit and allosteric changes dynamically
- Use of coarse-grained models to efficiently explore large conformational spaces

This allows for more realistic simulations of molecular interactions.

9.3 Improved solvent and environment modeling

Explicit inclusion of water molecules and ions in docking simulations is becoming more common, helping to better capture the role of the solvent environment in binding.

9.3.1 Hybrid multiscale modeling

The integration of docking with MD, QM/MM, and other computational chemistry methods will provide a more comprehensive view of molecular recognition and reaction mechanisms, bridging the gap between static docking poses and dynamic biological systems.

9.3.2 Expansion to novel targets and modalities

Docking is expanding beyond traditional small molecules to include:

- Peptides, proteins, and macrocycles
- Nucleic acids and complex biomolecular assemblies
- Covalent inhibitors and PROTACs (proteolysis-targeting chimeras)

This broadens the scope of therapeutic and biotechnological applications.

9.3.3 Cloud computing and high-performance resources

Access to powerful cloud-based platforms and GPUs allows the screening of ultra-large libraries with advanced methods, democratizing docking for researchers worldwide.

A comparison chart (**Figure 9**) illustrating differences between organic and green solvents used in bio-oil extraction, showing contrasts in toxicity, environmental impact, extraction efficiency, cost, and sustainability, with green solvents highlighted as safer and more eco-friendly alternatives. Organic solvents are widely used in bio-oil extraction due to their efficiency and availability. These solvents are typically petroleum-derived and are effective in dissolving a wide range of bioactive compounds. However, their use raises environmental and human health concerns, such as toxicity, flammability, and contribution to pollution.

On the other hand, green solvents represent a more sustainable alternative. Derived from bio-based or renewable sources, green solvents are designed to be environmentally friendly, biodegradable, and less toxic. They reduce the ecological footprint of bio-oil extraction and align with the principles of green chemistry.

Despite their advantages, the application of green solvents faces challenges and limitations. These include higher costs, limited availability, and sometimes lower extraction efficiency compared to traditional organic solvents. Research efforts are increasingly directed toward overcoming these challenges through the development of novel green solvent systems, process optimization, and large-scale feasibility studies.

The **Figure 9** highlights the trade-off between efficiency and sustainability in solvent selection, underlining the importance of future research and innovation to expand the applicability of green solvents in bio-oil extraction while minimizing environmental and health hazards.

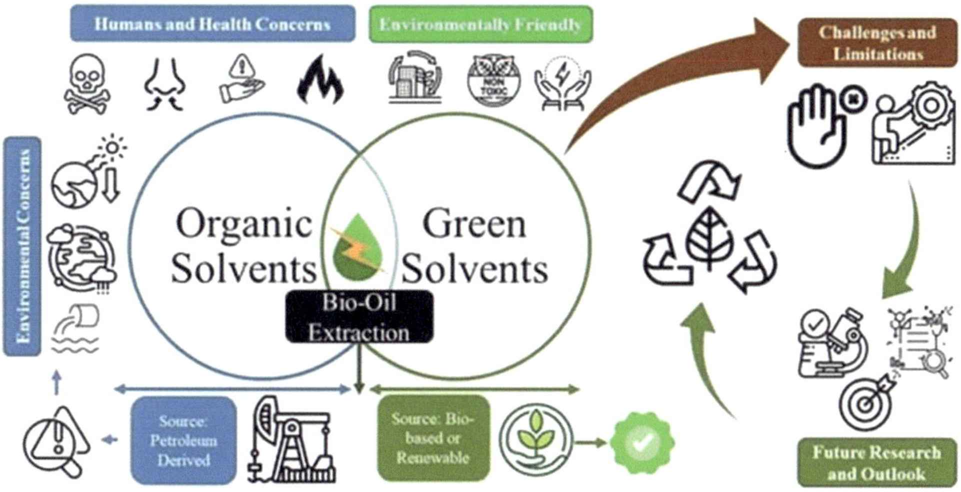

Figure 9.
Comparison of organic and green solvents in bio-oil extraction.

10. Conclusion

Molecular docking is a powerful computational technique that has revolutionized the fields of biomedical engineering and computational chemistry. By predicting how molecules interact at the atomic level, docking accelerates drug discovery, enzyme engineering, and biomolecular research. Despite existing challenges, such as flexibility treatment and scoring accuracy, ongoing advancements in algorithms, ML, and hybrid modeling approaches are continuously enhancing its reliability and scope. As computational resources expand and integration with experimental data deepens, molecular docking will play an increasingly vital role in designing next-generation therapeutics, biomaterials, and biochemical tools – paving the way for innovation in science and medicine.

An illustration, as shown in **Figure 10**, depicting molecular recognition in biological systems, showing how biomolecules such as proteins, DNA, and ligands interact through complementary shapes, charges, and binding sites to form specific and selective complexes. This **Figure 10** illustrates the specificity of molecular recognition in biological systems, showing how different biomolecules interact with high selectivity:

- **Antibody–antigen** → immune defense mechanism, where antibodies specifically bind to antigens

- **Enzyme–substrate** → catalysis, where enzymes recognize and process specific substrates

- **Transport–molecule** → selective transport of molecules (e.g., hemoglobin carrying oxygen)

- **Receptor–signal** → cell signaling, where receptors bind to specific ligands or signals to trigger responses

These examples demonstrate the "lock-and-key" principle, emphasizing how biological processes rely on precise molecular interactions for their function.

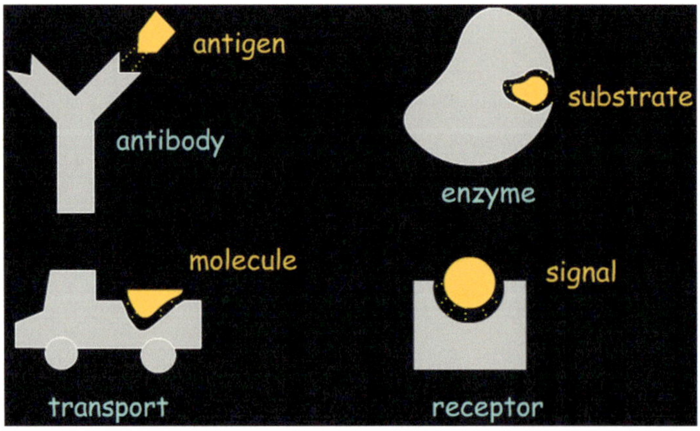

Figure 10.
Molecular recognition in biological systems.

DOI: http://dx.doi.org/10.5772/intechopen.1012645

Acknowledgments

I would like to express my heartfelt gratitude to the **Principal** and the **Management of Harsha Institute of Technology** for their constant encouragement and generous support throughout the preparation of this book chapter.

I am also thankful to my colleagues and peers, whose insights and discussions contributed to refining the ideas presented. Special thanks to my daughter, **Hithyshi K.**, an M.Tech. graduate in Computer Science & Engineering (AI), for her dedicated support in the development of this content.

I also appreciate the reviewers and editors for their valuable suggestions, which helped enhance the quality of the work. I would like to acknowledge the use of ChatGPT AI for its assistance in language refinement during the writing of this book chapter.

Finally, I acknowledge the indirect support from friends and family, whose patience and understanding made this academic contribution possible.

Author details

Bandenahalli Siddalingappa Krishna
Professor and HoD, Department of Chemistry, Harsha Institute of Technology, Nelamangala, India

*Address all correspondence to: drkrishnabs@gmail.com

References

[1] Wang R, Lai L, Wang S. Comparative evaluation of 11 scoring functions for molecular docking. Journal of Computational Chemistry. 2003;**24**:1672–1693

[2] Cheng T, Li X, Li Y, Liu Z, Wang R. Comparative assessment of scoring functions on a diverse test set (PDBbind). Journal of Chemical Information and Modeling. 2009;**49**:1079–1093

[3] Friesner RA, Banks JL, Murphy RB, et al. Glide: A new approach for rapid, accurate docking and scoring. Journal of Medicinal Chemistry. 2004;**47**: 1739–1749

[4] Cavasotto CN, Phatak SS. Homology modeling in drug discovery: current trends and applications. Drug Discovery Today. 2009;**14** (13–14):676–683. DOI: 10.1016/j.drudis.2009.05.010

[5] Rarey M, Kramer B, Lengauer T, Klebe G. A fast flexible docking method using an incremental construction algorithm (FlexX). Journal of Molecular Biology. 1996;**261**:470–489

[6] Ewing TJA, Makino S, Skillman AG, Kuntz ID. DOCK 4.0: Search strategies for automated molecular docking. Journal of Computer-Aided Molecular Design. 2001;**15**:411–428

[7] Ferreira LG, Dos Santos RN, Oliva G, Andricopulo AD. Molecular docking and structure-based drug design strategies. Molecules. 2015;**20**: 13384–13421

[8] Plewczynski D, Łaźniewski M, Augustyniak R, Ginalski K. Can we trust docking results? Evaluation on PD Bbind. Journal of Computational Chemistry. 2011;**32**:742–755

[9] Jones G, Willett P, Glen RC, et al. Development and validation of a genetic algorithm for flexible docking (GOLD). Journal of Molecular Biology. 1997;**267**:727–748

[10] Kuntz ID, Blaney JM, Oatley SJ, Langridge R, Ferrin TE. A geometric approach to macromolecule–ligand interactions. Journal of Molecular Biology. 1982;**161**:269–288

[11] Morris GM, Goodsell DS, Halliday RS, Huey R, Hart WE, Belew RK, Olson AJ. Automated docking using a Lamarckian genetic algorithm and an empirical binding free energy function. Journal of Computational Chemistry. 1998;**19**:1639–1662

[12] Kitchen DB, Decornez H, Furr JR, Bajorath J. Docking and scoring in virtual screening for drug discovery: methods and applications. Nature Reviews Drug Discovery. 2004;**3**(11):935–949. DOI: 10.1038/nrd1549

[13] Morris GM, Huey R, Lindstrom W, Sanner MF, Belew RK, Goodsell DS, Olson AJ. AutoDock4 and AutoDockTools4: Automated docking with selective receptor flexibility. Journal of Computational Chemistry. 2009;**30**(16):2785–2791. DOI: 10.1002/jcc.21256

[14] Trott O, Olson AJ. AutoDock Vina: improving the speed and accuracy of docking with a new scoring function, efficient

optimization, and multi-threading. Journal of Computational Chemistry. 2010;**31**(2):455–461. DOI: 10.1002/jcc.21334

[15] Halgren TA, Murphy RB, Friesner RA, Beard HS, Frye LL, Pollard WT, Banks JL. Glide XP: Accurate docking and scoring with a model of hydrophobic enclosure. Journal of Medicinal Chemistry. 2004;**47**:1750–1759

[16] Jones G, Willett P, Glen RC, Leach AR, Taylor R. Development and validation of a genetic algorithm for flexible docking (GOLD). Journal of Molecular Biology. 1997;**267**:727–748

[17] Liu Z, Li Y, Han L, et al. PDB-wide collection of binding affinities: Current status of the PDBbind database. Bioinformatics. 2015;**31**:405–412

[18] Sousa SF, Fernandes PA, Ramos MJ. Protein–ligand docking: current status and future challenges. Proteins: Structure, Function, and Bioinformatics. 2006;**65**(1):15–26. DOI: 10.1002/prot.21082

[19] Pierce BG, Hourai Y, Weng Z. Accelerating protein docking in ZDOCK using an advanced 3D convolution library. PLoS One. 2011;**6**(9):e24657. DOI: 10.1371/journal.pone.0024657

[20] Dominguez C, Boelens R, Bonvin AM. HADDOCK: a protein –protein docking approach based on biochemical or biophysical information. Journal of the American Chemical Society. 2003;**125**(7):1731–1737. DOI: 10.1021/ja026939x

[21] Friesner RA, Murphy RB, Repasky MP, Frye LL, Greenwood JR, Halgren TA, Sanschagrin PC, Mainz DT. Extra precision glide: Docking and scoring incorporating a model of hydrophobic enclosure for protein –ligand complexes. Journal of Medicinal Chemistry. 2006;**49**(21):6177–6196. DOI: 10.1021/jm051256o

Chapter 3

Harnessing Deep Learning and Generative AI for Molecular Docking Simulations: Tools, Challenges, and Future Directions

Bilal Shaker and Khaled Barakat

Abstract

Molecular docking has become a cornerstone in modern drug discovery, helping scientists predict how small molecules, or ligands, interact with target proteins. With the rise of artificial intelligence, particularly deep learning and generative models, traditional docking methods have been transformed. These AI-powered approaches not only boost accuracy but also dramatically reduce the time and cost of early-phase drug screening. This review explores how deep learning and generative AI are being applied to molecular docking simulations. Tools like AtomNet, DeepDock, KDEEP, and AutoDock Vina, when integrated with neural networks, have shown impressive improvements in predicting protein-ligand binding affinities compared to older docking methods. These AI models can rapidly screen massive chemical libraries, accelerating the identification of potential drug candidates. For instance, AtomNet, one of the first deep convolutional networks used in structure-based drug discovery, contributed to early drug leads for diseases like Ebola and cancer. AlphaFold, from DeepMind, has also revolutionized protein structure prediction, setting a new benchmark for accuracy. Beyond raw prediction power, these tools adapt over time. They learn from new data and minimizing human bias in compound selection. Their scalable architecture enables repeated simulations with increasing precision. This synergy of human intuition and AI precision marks a turning point in drug discovery. By using AI-driven molecular docking, researchers can now prioritize compounds more strategically, paving the way for faster, more efficient development of new treatments.

Keywords: deep learning, generative AI, molecular docking, drug discovery, protein-ligand interaction, deep generative models, virtual screening, scoring functions

1. Introduction

Drug discovery is a long, expensive, and complex process of identification and refinement of organic materials capable of interacting with a biological target, usually proteins. This is undertaken in vital steps with molecular docking. It is a computational technique that is employed to speculate the interaction of a small molecule (*i.e.,* ligand) with the target of interest (*i.e.,* protein). The relevance of molecular docking to structure-based drug design and virtual screening processes is involved when a researcher can narrow down the list of all possible drug candidates before investing on time and restricted financial resources in the costly laboratory tests [1].

Traditionally, molecular docking relies on search algorithms and scoring functions to predict the binding orientation and affinity of a ligand with the active site of a target protein. Despite their widespread use and their leading to many success stories, the traditional docking tools generally fail to handle the demands of the complexities of protein flexibility, solvent effects, and good scoring. Hence, they are susceptible to false-positive or opening of good molecules.

During the recent years, advances in artificial intelligence (AI) led to a phenomenal change in the field. To be more specific, deep learning (DL) and generative AI (GAI) are entering the domain of computational chemistry and drug designing, as game-changing technologies. These techniques provide new representation of molecules, the ability to learn based on existing information and the possibility to make predictions outside of the traditional hand-crafting of rules or energy calculations [2].

Deep learning is a type of machine learning algorithms, which learns complex patterns in data using artificial neural networks with many layers. Deep learning models are increasingly used in molecular docking to predict the binding affinity, classify docking poses, and improve the placement of protein-ligand complexes that often-outperforming traditional methods across various benchmark [3]. For example, conventional neural networks (CNNs) can treat protein-ligand complexes as 3D images, learning interactions patterns directly from structural data.

Generative AI aims at generating new molecular structures based on acquired distributions. Artificial intelligence is increasingly applied in *de novo* drug design through advanced generative models, including generative adversarial network (GANs), variational autoencoder (VAEs), and diffusion models. These approaches enable the exploration of vast chemical space and facilitate the generation of novel, drug-like compounds with the optimized properties tailored to specific targets or binding sites, often surpassing the limitations of existing chemical libraries and databases.

In combination, DL and GAN are presenting new opportunities in molecular docking simulations. With the employment of such AI methods into the field of drug discovery, scientists can attain:

a. More accurate prediction of binding affinities

b. Faster screening of millions of compounds

c. Better handling of molecular flexibility and chemical diversity

d. Automated generation of new leads for drug candidates.

Additionally, the docking tools based on AI that have demonstrated remarkable results in the recent research are referred to as GNINA [4], DiffDock [5], DeepDock [6], and AtomNet [7]. These methods are trained with big datasets of protein-ligand complexes, for example, those present in the PDBbind [8] and BindingDB databases [9], and learn to exploit real structural, experimental data.

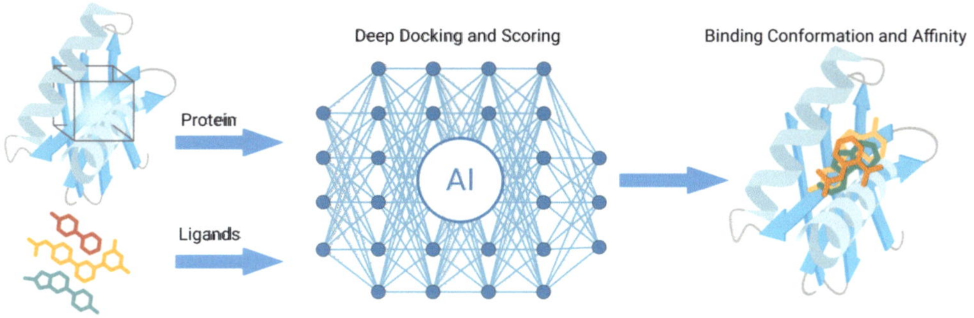

Figure 1.
Schematic representation of AI-assisted deep molecular docking. Protein structures and ligand molecules are input into deep learning frameworks, where advanced neural networks predict binding poses and estimate binding affinities. The integration of AI enables accurate identification of protein-ligand interactions, improving docking efficiency and reliability compared to traditional methods.

Regardless of the advances, significant challenges persist in the application of AI for drug discovery. These models often demand extensive, high-quality datasets and substantial computational resources, which can limit accessibility. Furthermore, issues of interpretability and generalizability remain unresolved, particularly when extending predictions to novel targets or underexplored protein families in the early stages of drug development. Conventionally, molecular docking uses scoring functions to rank ligands based on how well they might fit within a protein's binding site. Such classical techniques include AutoDock [10], Glide [11], and GOLD [12], whose discovery remains the subject of ongoing research. Besides, regulatory and ethical issues with using black-box AI models in healthcare-related decisions should be mitigated.

The current chapter will offer a broad picture of the use of deep learning and generative AI in molecular docking simulations (**Figure 1**). In the next sections we are going to discuss:

- Introduce the theoretical foundations of deep learning and generative models.
- Review the current state-of-the-art AI tools used in molecular docking.
- Analyze their development, architecture, and performance.
- Provide real-world case studies demonstrating their impact.
- Discuss the limitations, challenges, and potential future directions for integrating AI in drug discovery.

This is an attempt to provide a practical and accessible overview of this fast-growing area and thus ensure that it is of interest not just to computational chemists and AI researchers, but to biologists, pharmacologists, and to students wishing to gain insight into how the future of molecular docking may be like.

2. Overview of deep learning in molecular docking

In the last decade, numerous sectors such as computer vision, speech recognition, and natural language processing became revolutionized by deep learning (DL) [13].

Most recently, it has started revolutionizing life sciences, particularly fields such as genomics, structural biology, and drug discovery [13]. Deep learning helps drug discovery by locating, prioritizing, and verifying the ability of small molecules that can bind target proteins efficiently.

2.1 Introduction to deep learning methodology

Deep learning is a subfield of machine learning based on representations of artificial neural networks; computing systems inspired in part by human brain functions [14]. Such networks have layers of interconnected units (which are called neurons) that learn patterns of data by processing information. The characteristic that defines deep learning and makes it better than preceding machine learning techniques is the involvement of numerous layers (in many cases more than 3) that enable the system to be capable of learning higher abstractions of the information [14].

As an example, in image recognition edged recognition and color recognition may be possible features of early gross layers, whereas shapes, objects, or faces, those features learned by deeper fine layers. In molecular docking, similarly one might learn atom-level feature or type of bond in early layers, whereas deeper layers can learn spatial interactions or chemical environment or binding pattern.

2.2 Common deep learning architectures used in docking

Conventional docking applications are dependent on predefined scoring functions that are frequently based on empirical or physics-based guidelines designed to approximate protein-ligand binding affinities [13]. These techniques though serviceable to the field of drug discovery over decades are quite limited because of their inability to be generalized toward multiple protein targets and ligand chemotypes. Conversely, DL methods can be trained without predefinition on data, namely experimentally determined protein-ligand complexes, and can discover nonlinear and complex interaction patterns without the manual feature extraction or prior assumptions that characterize other methods.

Deep learning offers several key advantages over classical approaches:

- *Data-driven modeling:* DL eliminates the need to manually define interaction rules.
- *Broader generalization:* DL models can be trained across datasets with high chemical and structural diversity.
- *Enhanced accuracy:* Studies demonstrate improved binding affinity prediction over traditional scoring functions.
- *Task flexibility:* DL models are highly adaptable, being used for classification, regression, and pose prediction tasks.

Below are common DL architectures and the AI tools developed based on them:

2.2.1 Convolutional neural networks (CNNs)

Convolutional neural networks (CNNs) have been especially effective in working with structured 3D data, making them an ideal choice for protein-ligand docking tasks [15]. In molecular docking, both protein and ligand structures can be voxelized, which is the process of spatially representing the atoms and their chemical properties on a 3D grid. Each voxel represents some atom and encodes various molecular features including partial charges, hydrophobicity, and several other properties in a multi-channel format, similar to how images are stored in RGB channels.

CNNs focus on the spatially defined chemical environments of a protein's 3D grid using biologically motivated spatially aware convolutional filters to learn binding-critical chemical environments in an iterative and hierarchical fashion. Thus, networks can learn and model subtle spatial features like the hydrogen bonding potential, steric complementarity, and electrostatic interactions critical for binding poses and affinities prediction. The voxel-based architecture enables systematic molecular geometry handling, spatial orientation, and some degree of rotational invariance in a discretized form, which is important for the learning process of the CNN.

Many researches managed to use CNN for docking and scoring tasks. For example, in 2017, Ragoza et al. proposed a CNN model integrated on 3D convolutional neural networks which learned to predict binding directly [16].

GNINA represents a significant advancement in the field of molecular docking by integrating CNNs, which innovatively improves molecular docking workflows. Specifically, GNINA applies an AutoCNN scoring function, which is trained on the extensive CrossDocked2020 dataset. GNINA has proven its effectiveness by participating in numerous benchmarking contests, including the Drug Design Data Resource (D3R) Grand Challenges 4 and 5, consistently scoring as one of the better competing tools [15, 17].

2.2.2 Graph neural networks (GNNs)

Graph neural networks (GNN) are deep learning models that depict molecules in graph formats with atoms as nodes and chemical bonds as edges [18]. GNNs perform message passing through loops in which atom-level information is refined by information aggregation from adjacent atoms, allowing the model to capture intricate structural dependencies. Because of their unique architecture, GNNs are adept at learning the intricate spatial arrangements along with the electronic structure of molecules, improving performance in property prediction, estimating binding affinities, and molecular docking.

PotentialNet is a model based on a GNN that predicts the binding affinity of protein-ligand complexes as well as other molecular properties, leveraging structural information [19]. PotentialNet was trained on the PDBbind v2016 dataset, and, as of now, it is the most accurate model for these tasks. Moreover, it shows persistent performance for different chemical and biological domains. The architecture of PotentialNet is capable of integrating all covalent and non-covalent interactions *via* message passing on molecular graphs, which aids its performance in sophisticated biochemical domains. Most importantly, PotentialNet has been utilized in actual drug discovery projects, for example, in the lead compound identification in kinase inhibitor campaigns.

2.2.3 Recurrent neural networks (RNNs)

RNNs are special deep learning models that are good at processing sequential data. This makes them very effective at handling SMILES strings, which are text-based representations of molecular structures. RNNs are often used in generative chemistry to make new compounds, but they have also shown promise in tasks like virtual screening and ligand generation within molecular docking pipelines. Because they can capture sequential dependencies, they can model molecular features that are important for binding affinity and drug-likeness in a more detailed way.

A deep generative framework called Reinforcement Learning for Structural Evolution (ReLeaSE) uses reinforcement learning and RNNs to create and improve new drug-like compounds [20]. It has been effectively used to create ligands that target opioid receptors, demonstrating its potential for exploring and enlarging chemical space. It incorporates docking-based filters to prioritize compounds with predicted binding affinity.

2.3 Training data and benchmarks

Deep learning models for molecular docking depend heavily on large, high-standard datasets to ensure accurate and generalizable predictions. Several well-established databases serve as foundational sources for training and validating these models:

- *PDBbind*: Offers detailed protein-ligand complexes with experimentally measured binding affinities, enabling regression-based model training [8].

- *BindingDB*: Contains over a million protein-ligand binding interactions, supporting both classification and regression tasks [9].

- *ChEMBL*: A manually curated database of bioactive drug-like molecules along with their biological targets and assay data, widely used in predictive modeling [21].

To evaluate model performance, regression metrics such as root mean square error (RMSE) are used to assess how well a model can predict continuous values like binding affinities. For classification tasks, such as distinguishing binders from non-binders, models are evaluated using accuracy and area under the curve (AUC) from the receiver operating characteristic (ROC) analysis. These benchmarks ensure reliability and consistency in docking predictions across diverse datasets.

2.4 Limitations and challenges

Deep learning (DL) models continue to face a number of significant obstacles that limit their wider applicability in molecular docking and drug discovery, despite their rapid and potential advancements. Lack of data is one of the biggest problems, particularly for rare or poorly understood protein targets and ligand classes. There are gaps in coverage that affect generalizability to novel or less-studied targets because many datasets used to train DL models are biased toward well-characterized systems.

Overfitting, in which models become overly adapted to the training data and perform poorly on unseen examples, is another serious problem. As a result, the model's resilience and practicality are jeopardized. Concerns about interpretability are also raised by the fact that many DL models function as "black boxes," making predictions devoid of logical justification, which makes it challenging for researchers to validate or trust results.

Another obstacle is the computational cost. Not all research groups may have access to the significant computing resources needed for training and fine-tuning complex architectures, such as GPUs and large memory footprints.

Finally, the field is having more and more trouble with issues of reproducibility and openness. In order for DL to be a reliable scientific tool, other people must be able to repeat the experiments and get the same results. However, this process is often slowed down by the fact that source code, datasets, and detailed methods are not shared enough. To promote the reproducibility and reliability in AI-driven drug discovery, it is still important to make sure that tools are tested in experiments and that they are available to more people.

3. Generative AI techniques for molecular design

Generative artificial intelligence (generative AI or GAI) refers to a class of machine learning algorithms capable of producing new data that resembles the distribution of training data. In drug discovery, these models are revolutionizing molecular drug design by learning from known molecules and generating novel chemical structures with the desired pharmacological properties. When applied to molecular docking, generative AI not only proposes new ligands but can also optimize them for enhanced binding affinity, specificity, and drug-likeness—key characteristics in the development of effective therapeutics [22].

Unlike traditional virtual screening approaches constrained by predefined chemical libraries, generative AI models allow exploration of vast, uncharted regions of chemical space. When integrated with docking simulations, GAI frameworks form a feedback loop where generated molecules are evaluated, scored, and iteratively refined based on their predicted docking performance and biological relevance [23].

3.1 What makes generative AI different?

Generative AI differs fundamentally from predictive models, which are limited to estimating physicochemical properties of existing molecules. Instead, GAI is designed to generate entirely new molecular structures from scratch. This allows researchers to move beyond known chemical libraries and explore novel scaffolds, potentially leading to first-in-class drug candidates.

3.2 Core generative AI architectures in drug discovery

Generative artificial intelligence (GAI) has grasped the development of a new and effective method of *de novo* drug design which allows the design of brand-new chemical structures with enhanced biological properties (**Figure 2**).

Several key architectures underlie the success of generative models in drug design:

3.2.1 Variational autoencoders (VAEs)

Variational autoencoders (VAEs) are powerful generative models used to design novel molecules in drug discovery. They work by compressing a molecular structure, usually represented as a SMILES string or a molecular graph, into a compact numerical format known as a latent vector using an encoder [24]. A decoder then reconstructs the molecule from this vector. When trained on large compound datasets, VAEs learn to position chemically similar molecules close together in this latent space, forming a smooth and continuous "chemical manifold." This enables researchers to

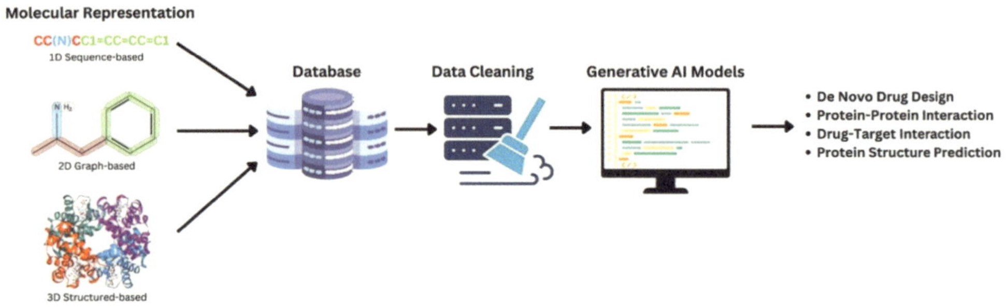

Figure 2.
Workflow of generative AI-driven molecular design and docking. Molecular representations (e.g., SMILES strings, molecular graphs, or protein structures) are first collected and organized into databases. Data curation ensures quality, consistency, and removal of redundant or noisy entries. Generative AI models (such as VAEs, GANs, transformers, or diffusion models) are then applied to learn molecular features and generative novel compounds. The generated molecules are evaluated for applications including de novo *drug design, molecular docking, and virtual screening, drug-target interaction prediction, and protein structure prediction.*

explore and interpolate between molecular structures, making it possible to generate entirely new compounds by sampling or modifying these latent vectors.

$$L_{VAE} = E_{(q^{\varnothing})\left(\frac{z}{x}\right)}\left[Inp_{è}\left(\frac{x}{z}\right)\right] - D_{KL}\left(q_{\varnothing}\left(\frac{z}{x}\right)\backslash P(z)\right) \quad (1)$$

where $(q^{\varnothing})\left(\frac{Z}{X}\right)$ is the encoder (approximate posterior), and $p_{è}\left(\frac{x}{z}\right)$ is the decoder (likelihood).

JT-VAE is a potent graph-based variational autoencoder that encodes molecules and their substructures as tree-like graphs in order to produce diverse and chemically valid molecular structures [25]. The model can create realistic and unique compounds while preserving structural consistency thanks to this chemically informed representation. JT-VAE proved its efficacy by correctly reconstructing 76% of ChEMBL test molecules and producing novel molecular scaffolds that passed synthetic accessibility tests after being trained on bioactive compound datasets like ZINC and ChEMBL. Because of these features, JT-VAE is a useful tool in the early stages of drug discovery, especially when investigating new chemical spaces that have characteristics similar to those of drugs.

3.2.2 Generative adversarial networks (GANs)

Generative adversarial networks (GANs) represent a distinctive approach to molecular generation by framing the task as a competitive game between two neural networks: a generator and a discriminator [22]. The discriminator determines whether the outputs are real (from the training set) or fake (generated), while the generator tries to create molecules—typically in the form of SMILES strings—from random input. This adversarial setup compels the generator to create increasingly realistic, drug-like molecules over time. Early applications in drug discovery, such as ORGAN by Insilico Medicine, used a recurrent neural network (RNN) as the generator and guided generation toward molecules with favorable properties using reward functions [26]. This framework is expanded by conditional GANs (cGANs), which uses extra input, like desired molecular properties, to direct generation toward particular therapeutic objectives [27].

Despite their promise, GANs face notable challenges when applied to molecular data, primarily due to the discrete nature of molecular representations, which complicates backpropagation. To overcome this, researchers have incorporated reinforcement learning strategies or adopted differentiable relaxations. For instance, Insilico's Adversarial Threshold Neural Computer merged GANs with reinforcement learning and a neural computer-based generator, allowing external rewards for properties like pharmacological activity and synthetic feasibility [28]. Similarly, MolGAN departed from sequence-based outputs and directly generated molecular graphs, achieving nearly 100% validity and enhanced synthetic accessibility on benchmark datasets such as QM9 [29]. Nonetheless, GANs are prone to issues like mode collapse, where the generator produces limited molecule types, and training instability. Yet, with careful design and conditional constraints, GANs remain a powerful tool for targeted molecule generation, offering significant potential in lead optimization and *de novo* drug discovery.

3.2.3 Transformer-based models

Transformer-based models, originally designed for natural language processing, have found powerful applications in molecular sciences due to their ability to capture long-range dependencies through self-attention mechanisms [30]. Transformers use chemical sequences as a language when applied to molecular strings such as SMILES or SELFIES. Models like ChemBERTa, trained on millions of SMILES strings using masked language modeling, generate rich molecular embeddings that can be used for property prediction, virtual screening, or molecule generation [31].

Large-scale protein language models, like ProGen, can *de novo* generate functionally active proteins by interpreting amino acid sequences as text [32]. Even with low sequence identity, ProGen has produced enzymes with catalytic activity that is strikingly similar to that of natural proteins, and X-ray crystallography has been used for experimental validation to confirm proper folding and functional sites. All things considered, transformer models help close the gap between sequence and structure by using self-supervised learning on large unlabeled datasets to generate structures, predict properties, optimize codons, and design analogs.

3.2.4 Denoising diffusion probabilistic models (DDPMs)

DDPMs represent a powerful new class of generative models that produce data by reversing a gradual noise-injection process [33]. These models begin with pure Gaussian noise and iteratively refine it using a trained neural network, reconstructing realistic samples such as 3D molecular structures or graphs. Unlike VAEs or GANs that operate in latent space, diffusion models work directly in the data space, enabling generation of complex, high-fidelity structures like protein-ligand poses. This methodology is applied to molecular docking and protein design by models such as DiffDock [5] and RFdiffusion [34], which produce the state-of-the-art results in pose prediction and binder generation with remarkable structural and functional accuracy.

Despite their strength, diffusion models are computationally intensive, often requiring thousands of iterative steps. To address this, faster variants like DDIMs and distilled models such as SGMs have been developed, reducing inference time dramatically with little loss in quality. Cascade and hierarchical models further break down the generation into coarse and fine steps to balance efficiency and resolution. These advances have enabled real-world applications, such as designing novel proteins and enzymes with confirmed binding affinities and catalytic activities, and validated through experiments and structural biology. Diffusion models, alongside VAEs, GANs, and transformers, are now a key part of modern generative pipelines in drug discovery—frequently used in hybrid workflows to combine their strengths in accuracy, diversity, and interpretability.

3.2.5 Comparative landscape of generative models in molecular design

Generative models such as VAEs, GANs, DDPMs each bring unique advantages and limitations to molecular generation tasks. With models like transformer–VAE hybrids attaining up to 90% reconstruction accuracy and grammar—or graph-based extensions (e.g., MHG–VAE) reaching 100% validity, VAEs are preferred for their stable training dynamics and interpretable latent spaces. They are particularly effective in analog generation and latent space optimization but may struggle with molecular validity and direct control over properties when not finely tuned.

Conversely, GANs provide conditional control for synthesis that targets specific properties and are excellent at producing chemically realistic molecules [22]. However, these models often suffer from training instability and mode collapse, reported in over 30% of reinforcement learning augmented runs. This makes them less reliable for large scale chemical exploration unless stabilized with techniques such as Wasserstein loss or actor-critic frame works. These models are more suited to focused lead optimization tasks. Transformer-based models, like ChemBERTa trained on approximately 77 million SMILES strings, have demonstrated high performance in property prediction and robust molecular embeddings across MoleculeNet benchmarks. Although they are computationally intensive and necessitate thorough chemical validity checks after generation, they are ideal for sequence-based generation and representation learning.

Finally, DDPMs have emerged as the state of the art for high-fidelity molecular and structural generation. Tools such as DiffDock can function ~40× faster with DDIM acceleration and have demonstrated remarkable pose prediction accuracy (median RMSD ~1.5 Å, >75% success under 2 Å). Nonetheless, standard DDPMs remain computationally intensive, often requiring 1000–4000 inference steps per sample. Recent advances such as DDIMs, progressive distillation, and cascade diffusion have significantly reduced this burden [35].

In conclusion, the choice of a model should be in line with particular research objectives: Transformers are ideal for molecular language modeling and property embedding, VAEs and GANs are best for 2D scaffold generation and analog design, and DDPMs are best for geometry-aware, structure-based applications like protein-ligand docking (**Table 1**).

3.3 Challenges and opportunities in generative AI for drug discovery

Generative AI has rapidly advanced molecular design and drug discovery by enabling the creation of novel chemical structures beyond existing libraries. However, a number of obstacles still stand in the way of its widespread adoption. A major challenge is synthetic feasibility—many AI-generated molecules, though theoretically valid, are impractical or impossible to synthesize in real-world lab settings due to complex or unstable structures. Another significant issue is data bias, where training datasets, such as ZINC or ChEMBL, inherently favor certain chemical scaffolds or pharmacophores, leading to limited chemical diversity in generated outputs. Moreover, evaluation of generated molecules remains an open problem.

Datasets / Benchmarks	**ZINC, QM9**	**QM9, ZINC**	**UniProt, ZINC, PDB**	**PDBbind, PoseBusters**
Performance Highlights	Validity: 60–98%; High uniqueness; Moderate synthetic accessibility (SA) scores	Validity up to 100% (MolGAN); High SA scores; Moderate molecular diversity	Low perplexity; ~99% validity; 20–40% sequence identity in protein generation	Pose RMSD (DiffDock): ~1.5 Å; Validity ~100%; >75% success rate in pose prediction
Key Tools / Examples	Junction Tree VAE, Grammar VAE, GSchNet	ORGAN, MolGAN, Adversarial Threshold Neural Computer	ChemBERTa, ProGen, GPT-chem, TAPE	GeoDiff, RFdiffusion, DiffDock
Primary Use Case	Small molecule generation	Small molecule design	Molecular and protein modeling	3D molecular generation and docking
Typical Molecular Representations	SMILES, molecular graphs, 3D coordinates	SMILES, molecular graphs	SMILES, SELFIES, protein sequences	3D coordinates, molecular graphs, atom matrices
Challenges / Limitations	May generate invalid structures; latent variables may not directly correlate with properties.	Susceptible to training instability and mode collapse; non-differentiable outputs for molecular graphs.	High computational demand; requires large training corpora; indirect control over molecular properties.	Generation can be computationally slow (many denoising steps); relatively new in chemistry domain.
Advantages	Smooth, interpolatable latent space; stable training; Bayesian interpretability.	Produces realistic and diverse outputs; supports conditional generation; compatible with reinforcement learning.	Effective at capturing long-range dependencies; excellent for language-like sequence modeling; enables pretraining on large datasets.	Highly diverse and high-quality outputs; handles 3D data effectively; theoretically complete modeling.
Core Concept	Encode molecules into a continuous latent space and decode to reconstruct them, optimizing the Evidence Lower Bound (ELBO).	Use a generator-discriminator framework trained adversarially to mimic real molecular distributions.	Utilize self-attention mechanisms to model dependencies across molecular or protein sequences.	Generate molecules by learning to reverse a gradual noise process applied to input data.
Model Type	Variational Autoencoders (VAEs)	Generative Adversarial Networks (GANs)	Transformer-Based Models	Denoising Diffusion Probabilistic Models (DDPMs)

Table 1.
Comparative summary of deep generative model architectures in molecular design.

While metrics such as validity, uniqueness, and QED scores offer insights, they do not guarantee biological relevance or activity, raising questions about how much researchers can trust AI-generated compounds without experimental validation.

Furthermore, there is increasing concern in the field regarding the interpretability of generative models. The majority of deep learning architectures, such as transformers, GANs, and VAEs, function as "black boxes," offering little explanation for the rationale behind why a specific molecule was proposed or how it meets the design criteria. This lack of transparency can hinder decision-making in critical applications like lead optimization or safety assessment. Future studies are shifting toward hybrid modeling techniques to address these problems, which blend the rigorous predictive accuracy of physics-based models like molecular docking and quantum mechanics with the creative potential of generative AI. These hybrid frameworks aim to balance innovation and realism—generating molecules that are not only novel and bioactive but also synthetically accessible and explainable. Some notable approaches include integrating reinforcement learning with docking scores as feedback signals (e.g., DeepLigand) or embedding quantum mechanical descriptors within generative workflows (e.g., PhysNet). These techniques present encouraging avenues for improving AI-driven drug discovery's dependability and usefulness [23, 36].

4. AI-powered tools for molecular docking: Features, strengths, success stories, and shortcomings

The emergence of artificial intelligence has created new docking tools, which uses deep learning and generative modeling, to integrate in the structure-based drug discovery pipelines. These tools are designed to overcome fundamental shortcomings of classical docking (improper scoring functions, inflexible protein model, and lack of generalization), by using training on data. Within this section, we discuss the validity of some of the state of the art of AI-powered docking platforms, evaluate their methods, and determine the contributions and weaknesses of the platforms in the field of molecular simulation.

4.1 GINNA

GNINA (pronounced "genie-nuh") is a hybrid framework that enhances traditional docking by integrating 3D convolutional neural networks (CNNs) with the AutoDock Vina engine. Unlike classical scoring methods based on physics or empirical rules, GNINA uses deep learning to evaluate protein-ligand interactions. It converts molecular complexes into 3D voxel grids—where atom types and their spatial arrangements are encoded—and processes them through CNNs to predict binding poses and affinities more accurately. Trained on large datasets like PDBbind, GNINA delivers improved performance in docking predictions.

GNINA calculates a binding score using the following CNN-based regression model:

$$\hat{y} = f_\theta(V) \tag{2}$$

Where $\hat{y}$ is the predicted binding score (affinity or pose confidence), V is the 3D voxel grid representation of the protein-ligand complex, and f_θ is the CNN model with learnable parameters θ.

This score is often combined with the traditional Vina scoring function:

$$S_{GNINA} = \propto . S_{CNN} + \beta . S_{Vina} \tag{3}$$

Where S_{CNN} is the CNN-predicted score, S_{Vina} is the AutoDock Vina empirical score, and α, β are weighting coefficients adjusted during training or tuning.

GNINA is trained using the PDBbind refined and general datasets, which provide high-quality protein-ligand crystal structures along with experimentally measured binding affinities. The model is optimized to minimize the mean squared error (MSE) between the predicted and actual binding affinities. Additionally, GNINA is trained to accurately identify the correct binding pose of ligands within the target protein site, enhancing both prediction accuracy and structural relevance in docking simulations.

One of the most positive achievements of GNINA is that it passed the D3R Grand Challenge, an evaluation event in structure-based drug design. In a number of rounds, GNINA overperformed numerous conventional and hybrid docking programs in terms of precision in pose prediction and accuracy to rank binding affinity. As one example, GNINA had the highest level performance as analyzed in the D3R GC4 challenge, where it correctly predicted ligand binding poses of inhibitors of BACE1. There is also the case that GNINA has been applied effectively in real-world applications to pharmaceutical pipelines to screen millions of compounds in fragment-based drug discovery or hit-to-lead optimization to help provide reduced false positives and faster candidate selection [17].

4.2 AtomNet

AtomNet groundbreaking DL model is one of the first and most important deep convolutional neural networks made for structure-based drug design. It used a new method by showing protein-ligand complexes as 3D voxel grids, which are basically small 3D pixels that show how atoms are arranged in space and what types they are. This helped the model learn how proteins and possible drug molecules interact with each other in complex ways in space.

Virtual screening efforts, particularly in identifying lead compounds for challenging targets such as Ebola virus proteins and cancer-related enzymes, AtomNet's predictions have guided experimental validation, uncovering molecules with strong potential as therapeutic leads [37].

The success of AtomNet marked a turning point in computational drug discovery, demonstrating that deep learning can outperform traditional docking methods in pose prediction and virtual screening accuracy. As the first deep learning model to achieve such performance in structure-based design, AtomNet laid the foundation for a new generation of AI-driven drug discovery tools.

4.3 DiffDock

DiffDock presents a new generative paradigm to docking, based on the diffusion models that are trained not to search but to generate the binding poses [5]. It radically contrasts with the traditional docking methods, according to which generation and ranking of poses rely much on scoring functions and optimization heuristics. DiffDock, on the other hand, learns how to sample realistic poses guided by protein pocket geometries and ligand structure.

DiffDock is surprisingly effective at generalization tasks and a more flexible protein-ligand interaction model. It does well on unseen targets, and it permits more biologically realistic binding conformations. But it is time-consuming, and it has not been validated against a larger diversity of proteins and chemical scaffolds yet.

Compared to the classical docking tools, where there was difficulty in estimation of the binding poses in more than two kinases, DiffDock could predict realistically the ligand conformations with only a fraction of the cost in computation time. These poses perfectly aligned with X-ray crystal structures (RMSD <2.0 A) and directed the further optimization of the lead. DiffDock offered a shortcut around the conventional scoring and search procedures to allow medicinal chemists quicker feedback in the design-make-test cycle.

4.4 DeepDock

DeepDock uses deep learning in order to optimize docking scores using graph-based modeling. DeepDock represents molecules as molecular graphs rather than grids and processes these molecular representations with GNNs to learn about atomic interactions in terms of relationships. The tool does that by training on binding affinities in the real world and enhancing the scoring accuracy.

The compatibility of DeepDock with benchmark datasets such as the CASF-2016 benchmark [38], where DeepDock performs significantly better than classical scoring functions, is also part of the strengths of this computing tool. The tool might, however, need a lot of preprocessing and cannot handle more complicated forms of dynamics including water mediation and protein conformational changes.

DeepDock is used in a number of academic organizations in lead discovery. In one of the European universities, scientists had been working on the development of matrix metalloproteinase (MMPs) inhibitors a set of enzymes present in body tissues and known to be involved in tissue breakdown and tumor spreading.

With DeepDock graph-based scoring model, the group could filter large docking campaigns output and eliminate false positives. The scoring model was able to characterize the most important molecular interactions than the standard empirical functions, as indicated by follow-up biological testing reaffirming the nanomolar action of several of the highest ranked compounds.

5. Challenges and limitations of AI in molecular docking

The integration of artificial intelligence into molecular docking has transformed drug discovery by improving pose prediction, virtual screening, and molecular generation. However, despite these advancements, several limitations hinder its full potential. Understanding these challenges is crucial for refining AI tools and ensuring their applicability in real-world pharmaceutical research.

The strong dependence on data quality is one of the main issues. Many databases, including PDBbind and BindingDB, are biased toward well-studied protein families like kinases or GPCRs, and deep learning models are only as good as the datasets they are trained on. This bias can lead to models that perform well on familiar targets but fail to generalize to novel proteins or rare molecular scaffolds. Furthermore, model accuracy may be impacted by noise introduced by variations in experimental protocols and dataset resolution.

Another major issue is the lack of interpretability. Unlike traditional docking tools that produce physically explainable outputs (e.g., hydrogen bonds or binding energies), many AI models function as "black boxes." This opacity hinders trust, reproducibility, and regulatory approval. Although techniques like attention maps and saliency scores are being explored for model explainability, they remain underdeveloped and not widely adopted.

Another obstacle is super computational demands. Large GPU resources are needed to train state-of-the-art models like diffusion-based networks or 3D CNNs, which limits their applicability in settings with limited resources or in smaller institutions. Furthermore, many models scale poorly during inference, particularly when predicting docking for large libraries or flexible binding sites. Finally, challenges such as poor generalization to novel targets, the absence of standardized evaluation metrics, ethical concerns, and limited integration with experimental validation further highlight the need for robust, transparent, and collaborative AI-driven drug discovery frameworks.

6. Future directions and opportunities for AI in molecular docking

As artificial intelligence (AI) continues to evolve, its role in molecular docking is set to become a foundational aspect of drug discovery. Beyond improving the accuracy of predictions, future advancements in AI-driven docking will emphasize transparency, adaptability, and broader applicability. One of the most pressing needs is enhancing model interpretability. Scientists and regulatory bodies alike are increasingly expecting AI tools to explain their predictions. Techniques such as attention mechanisms, saliency maps, and interaction heatmaps are being explored to clarify which parts of a molecule contribute most to binding affinity. This transparency could help medicinal chemists make informed decisions rather than relying solely on AI-generated rankings.

Another exciting direction is the integration of multi-modal and hybrid modeling approaches. Future AI systems are likely to combine structural data with insights from molecular dynamics, protein–protein interactions, gene expression, and even patient-specific information. These hybrid models could make highly contextual predictions, such as how a compound might behave in a particular cellular environment or interact with off-target proteins. Moreover, combining AI with physics-based simulations can offer the best of both worlds: accuracy and interpretability.

Generative models will also play a transformative role in ligand design and optimization. Emerging tools, such as diffusion models and large-scale chemical language models, promise to generate complex 3D molecular structures, selectivity filters, and pharmacophore maps in real time. These tools are expected to enable dynamic lead optimization processes by integrating docking, scoring, and experimental planning in a continuous feedback loop. Additionally, strategies like transfer learning and few-shot learning will help extend these tools to rare diseases and novel targets where large datasets are unavailable.

Equally important is the movement toward greater accessibility and collaboration. Cloud-based AI-as-a-service platforms are making powerful models available to researchers worldwide, leveling the playing field between resource-rich and smaller labs. Future platforms may facilitate interdisciplinary collaborations, allowing scientists from various domains to work together in real time. To support all this progress, ethical standards and regulatory frameworks must also evolve. Transparent documentation, model versioning, and reproducibility will be essential to building trust in AI-based discoveries. In sum, the future of AI in molecular docking is not just about better predictions—it is about transforming the entire scientific process through openness, collaboration, and smart integration.

7. Conclusion

Artificial intelligence has become the revolution in the world of molecular docking by changing the course of drug discovery. Since ancient times of rule-based docking algorithms, up to the state of the art of deep learning and generative algorithms, the history of AI has managed to solve longstanding issues of accuracy, speed, and scaling. With neural networks through machine learning-enabled scoring functions, pose prediction, and AI-aided ligand generation, the researchers can access tools that not only are faster but are also smarter.

The inclusion of AI has greatly increased the speed at which the preliminary phases of drug design are achieved in that it can screen huge chemical libraries virtually, calculate predicted binding affinities using refined accuracy, and optimize the structures of ligands *in silico* prior to actually synthesizing them. Notably, AI models have also introduced a paradigm shift to data-driven discovery, that is, predictive insights are used to prioritize which insights will be tested experimentally, lowering the time and cost throughout the development pipeline.

Though such improvements have been made, challenges are still encountered. The problems of model interpretability, data quality, and generalizability should be handled to provide reliable predictions made by AI. Ethics and governance should also be brought into play to help determine the responsible application of AI technologies, especially since they are being mainstreamed as part high-stakes human health applications.

The future of the marriage of AI and molecular docking is not an understatement. New technologies like explainable AI, multi-modal integration, transfer learning, and cloud-based systems will further standardize access and increase the accuracy of the drug design tools. As computational scientists, biologists, and clinicians continue to work together, AI will not only become an assistant but a fundamental pillar in the pharmaceutical studies of a new generation.

In this evolving landscape, explainable AI and reproducibility have become central concerns. Even though AI-driven docking can produce remarkably accurate results, many deep learning models' decision-making processes are still opaque, which undermines regulatory acceptance and scientific confidence. The goal of explainable AI techniques like saliency maps, interaction heatmaps, and attention mechanisms is to make it clear which molecular characteristics influence predictions so that biologists and chemists can better understand the findings.

Reproducibility is equally important, although independent validation is hampered by the fact that many AI models rely on proprietary datasets, non-standardized benchmarks, or opaque training pipelines. To guarantee that AI-generated predictions can be relied upon, duplicated, and incorporated into actual drug discovery processes, it is imperative to establish open datasets, open reporting procedures, and community-wide benchmarking challenges (such as CASF and D3R).

Overall, AI is not replacing human factor in drug discovery, and on the contrary, it is improving it. AI means a future in which effective therapies are found sooner, more safely, and with more certainty: using computational power to augment human intuition. Where innovation, ethics, and interdisciplinary science intersect, there will be the extent to which this transformation can reach across and how many lives can benefit in the end.

Acknowledgements

K.B would like to acknowledge funding from the Canadian Institutes of Health Research, Kidney Foundation, Li Ka Shing Institute of Virology and Natural Sciences and Engineering Research Council.

Conflict of interest

The authors declare that they have no conflict of interest.

Author details

Bilal Shaker[1*] and Khaled Barakat[1,2,3*]

1 Faculty of Pharmacy and Pharmaceutical Sciences, University of Alberta, Edmonton, AB, Canada

2 Li Ka Shing Institute of Virology, University of Alberta, Edmonton, AB, Canada

3 Thoth Biosimulations Inc, Edmonton, AB, Canada

*Address all correspondence to: bshaker@ualberta.ca and kbarakat@ualberta.ca

References

[1] Ain QU, Aleksandrova A, Roessler FD, Ballester PJ. Machine-learning scoring functions to improve structure-based binding affinity prediction and virtual screening. Wiley Interdisciplinary Reviews: Computational Molecular Science. 2015;**5**:405-424

[2] Ballester PJ, Mitchell JBO. A machine learning approach to predicting protein-ligand binding affinity with applications to molecular docking. Bioinformatics. 2010;**26**(9):1169-1175

[3] Brown T, Mann B, Ryder N, Subbiah M, Kaplan JD, Dhariwal P, et al. Language models are few-shot learners. Advances in Neural Information Processing Systems. 2020;**33**:1877-1901. [Internet]. Available from: https://commoncrawl.org/the-data/

[4] McNutt AT, Francoeur P, Aggarwal R, Masuda T, Meli R, Ragoza M, et al. GNINA 1.0: Molecular docking with deep learning. Journal of Cheminformatics. 2021;**13**(1):43

[5] Corso G, Stärk H, Jing B, Barzilay R, Jaakkola T. DiffDock: Diffusion steps, twists, and turns for molecular docking. Preprint at arXiv. In: International Conference on Learning Representations (ICLR). 2023. Available from: http://arxiv.org/abs/2210.01776

[6] Gentile F, Agrawal V, Hsing M, Ton AT, Ban F, Norinder U, et al. Deep docking: A deep learning platform for augmentation of structure based drug discovery. ACS Central Science. 2020;**6**(6):939-949

[7] Yu J, Wang Z, Saksena A, Wei S, Wei Y, Colnaghi T, et al. 3D deep learning for enhanced atom probe tomography analysis of nanoscale microstructures. Acta Materialia. 2024;**278**:120280

[8] Wang R, Fang X, Lu Y, Yang CY, Wang S. The PDBbind database: Methodologies and updates. Journal of Medicinal Chemistry. 2005;**48**(12):4111-4119

[9] Liu T, Lin Y, Wen X, Jorissen RN, Gilson MK. BindingDB: A web-accessible database of experimentally determined protein-ligand binding affinities. Nucleic Acids Research. 2007;**35**:D198-201

[10] Gaillard T. Evaluation of AutoDock and AutoDock Vina on the CASF-2013 benchmark. Journal of Chemical Information and Modeling. 2018;**58**(8):1697-1706

[11] Halgren TA, Murphy RB, Friesner RA, Beard HS, Frye LL, Pollard WT, et al. Glide: A new approach for rapid, accurate docking and scoring. 2. Enrichment factors in database screening. Journal of Medicinal Chemistry. 2004;**47**(7):1750-1759

[12] Verdonk ML, Cole JC, Hartshorn MJ, Murray CW, Taylor RD. Improved protein-ligand docking using GOLD. Proteins: Structure, Function and Genetics. 2003;**52**(4):609-623

[13] Askr H, Elgeldawi E, Aboul Ella H, Elshaier YAMM, Gomaa MM, Hassanien AE. Deep learning in drug discovery: An integrative review and future challenges. Artificial Intelligence Review. 2023;**56**(7):5975-6037

[14] Sharifani K, Amini M. Machine learning and deep learning: A review of methods and applications. World Information Technology and Engineering Journal. 2023;**10**(07):3897-3904. Available from: https://ssrn.com/abstract=4458723

[15] Francoeur PG, Masuda T, Sunseri J, Jia A, Iovanisci RB, Snyder I, et al. Three-dimensional convolutional neural

networks and a crossdocked data set for structure-based drug design. Journal of Chemical Information and Modeling. 2020;**60**(9):4200-4215

[16] Ragoza M, Hochuli J, Idrobo E, Sunseri J, Koes DR. Protein-ligand scoring with convolutional neural networks. Journal of Chemical Information and Modeling. 2017;**57**(4):942-957

[17] Koukos PI, Xue LC, Bonvin AMJJ. Protein–ligand pose and affinity prediction: Lessons from D3R grand challenge 3. Journal of Computer-Aided Molecular Design. 2019;**33**(1):83-91

[18] Xiong J, Xiong Z, Chen K, Jiang H, Zheng M. Graph neural networks for automated de novo drug design. Drug Discovery Today. 2021;**26**(6):1382-1393

[19] Feinberg EN, Sur D, Wu Z, Husic BE, Mai H, Li Y, et al. PotentialNet for molecular property prediction. ACS Central Science. 2018;**4**(11):1520-1530

[20] Agyemang B, Wu WP, Addo D, Kpiebaareh MY, Nanor E, Roland HC. Deep inverse reinforcement learning for structural evolution of small molecules. Briefings in Bioinformatics. 2021;**22**(4):bbaa364

[21] Gaulton A, Hersey A, Nowotka ML, Patricia Bento A, Chambers J, Mendez D, et al. The ChEMBL database in 2017. Nucleic Acids Research. 2017;**45**(D1):D945-D954

[22] Kanakala GC, Devata S, Chatterjee P, Priyakumar UD. Generative artificial intelligence for small molecule drug design. Current Opinion in Biotechnology. 2024;**89**:103175

[23] Zhavoronkov A, Ivanenkov YA, Aliper A, Veselov MS, Aladinskiy VA, Aladinskaya AV, et al. Deep learning enables rapid identification of potent DDR1 kinase inhibitors. Nature Biotechnology. 2019;**37**(9):1038-1040

[24] Ochiai T, Inukai T, Akiyama M, Furui K, Ohue M, Matsumori N, et al. Variational autoencoder-based chemical latent space for large molecular structures with 3D complexity. Communications Chemistry. 2023;**6**(1):249

[25] Gonchar AV, Tuzikov AV, Andrianov AM, Furs KV. Application of semi-supervised GAN in combination with JT-VAE for generation of small molecules with high binding affinity to the KasA enzyme of Mycobacterium tuberculosis. In: Proceedings of the 16th International Conference, United Institute of Informatics Problems of the National Academy of Sciences of Belarus. Minsk; 2023. pp. 114-117. [Internet]. Available from: https://www.rcsb.org/structure/4c72

[26] Tripathi S, Augustin AI, Dunlop A, Sukumaran R, Dheer S, Zavalny A, et al. Recent advances and application of generative adversarial networks in drug discovery, development, and targeting. Artificial Intelligence in the Life Sciences. 2022;**2**

[27] Kucera T, Togninalli M, Meng-Papaxanthos L. Conditional generative modeling for de novo protein design with hierarchical functions. Bioinformatics. 2022;**38**(13):3454-3461. Available from: https://academic.oup.com/bioinformatics/article/38/13/3454/6593486

[28] Putin E, Asadulaev A, Vanhaelen Q, Ivanenkov Y, Aladinskaya AV, Aliper A, et al. Adversarial threshold neural Computer for Molecular de novo design. Molecular Pharmaceutics. 2018;**15**(10):4386-4397

[29] De Cao N, Kipf T. MolGAN: An implicit generative model for

small molecular graphs. preprint arXiv:2022:11973. Available from: http://arxiv.org/abs/1805.11973

[30] Jiang J, Ke L, Chen L, Dou B, Zhu Y, Liu J, et al. Transformer technology in molecular science. Wiley Interdisciplinary Reviews: Computational Molecular Science. 2024;**14**(4):e1725

[31] Mswahili ME, Jeong YS. Transformer-based models for chemical SMILES representation: A comprehensive literature review. Heliyon. 2024;**10**:e39038

[32] Gong M, Chen L, Li J. ProGen: Revisiting Probabilistic Spatial-Temporal Time Series Forecasting from a Continuous Generative Perspective Using Stochastic Differential Equations. preprint arXiv: arXiv:2024:2411.01267. Available from: http://arxiv.org/abs/2411.01267

[33] Parigi M, Martina S, Caruso F. Quantum-noise-driven generative diffusion models. Advanced Quantum Technologies. 2024;**15**:2300401

[34] Watson JL, Juergens D, Bennett NR, Trippe BL, Yim J, Eisenach HE, et al. De novo design of protein structure and function with RF diffusion. Nature. 2023;**620**(7976):1089-1100

[35] Aggarwal R, Koes DR. Learning RMSD to Improve Protein-Ligand Scoring and Pose Selection. ChemRxiv. 2020. DOI: 10.26434/chemrxiv.11910870.v2

[36] Gao W, Coley CW. The synthesizability of molecules proposed by generative models. Journal of Chemical Information and Modeling. 2020;**60**(12):5714-5723

[37] Kwofie SK, Adams J, Broni E, Enninful KS, Agoni C, Soliman MES, et al. Artificial Intelligence, machine learning, and big data for Ebola virus drug discovery. Pharmaceuticals. 2023;**16**(3):332

[38] Li Y, Su M, Liu Z, Li J, Liu J, Han L, et al. Assessing protein-ligand interaction scoring functions with the CASF-2013 benchmark. Nature Protocols. 2018;**13**(4):666-680

Chapter 4

Docking-Guided Discovery of Neuromodulatory Agents for Brain Cancer

Monochura Saha, Shubham Yadav, Habiba Zeb and Ishaq N. Khan

Abstract

High-grade gliomas exploit neural circuits to support growth, creating neuromodulatory vulnerabilities amenable to pharmacological targeting. This chapter presents a practical pipeline that centers on molecular docking and complementary in-silico methods to discover and prioritize neuromodulatory compounds for brain cancer. We summarize neuron-glioma signaling targets (glutamatergic/GABAergic receptors, synaptogenic proteins, ion channels), outline virtual screening and flexible docking approaches, and describe follow-up stability assessments with molecular dynamics (MD), artificial intelligence/machine learning (AI/ML) prioritization, and network-pharmacology mapping. Representative examples, such as perampanel, gabapentin, and valproic acid, illustrate opportunities and translational barriers, particularly blood–brain barrier penetration, protein flexibility, and scoring limitations. Finally, we propose experimental validation paths and delivery strategies, such as focused ultrasound and nanoparticle systems, to bridge computational predictions and clinical translation. Emphasizing rigorous docking methodology and integrated validation will accelerate the repurposing and discovery of neuromodulatory agents for brain cancer.

Keywords: molecular docking, cancer neuroscience, neuron–glioma synapse, neuromodulatory therapy, drug repurposing, blood–brain barrier, focused ultrasound, nanoparticle delivery, molecular dynamics, AI/ML-driven drug discovery, synaptogenic proteins, ion channels

1. Introduction

Brain tumors, particularly high-grade gliomas such as glioblastoma multiforme (GBM), present a formidable challenge in oncology. These neoplasms are characterized by aggressive growth, highly infiltrative behavior, and a dismal prognosis. Despite aggressive multimodal therapy, patients with glioblastoma have a median survival of approximately 15 months (versus roughly 3–4 months without treatment) [1]. The five-year survival rate remains poor, at about 4.7% [2]. The inherent heterogeneity of these tumors, manifesting both across different patients (inter-tumoral) and within a single tumor (intra-tumoral), significantly contributes to therapeutic resistance and frequent recurrence, undermining the efficacy of current treatment paradigms [3]. Conventional therapeutic approaches for brain tumors typically involve maximal safe surgical resection, followed by radiation therapy and chemotherapy, often utilizing agents like temozolomide [1]. However, these established modalities frequently encounter limitations, including significant side effects and the tumor's capacity to develop resistance, which collectively restrict their long-term effectiveness [4]. The persistent challenge of brain cancer, despite multimodal conventional therapies, underscores a critical need for innovative therapeutic strategies. This situation necessitates a deeper understanding of the unique biological vulnerabilities of brain tumors, especially their intricate interactions with the neural microenvironment, and a corresponding shift toward targeting these newly identified vulnerabilities.

In response to these persistent challenges, neuromodulation is emerging as a promising frontier in oncology, extending its application beyond traditional roles in pain management and neurological disorders [5]. Neuromodulation, broadly defined, involves regulating nervous activity either by controlling the physiological levels of neurotransmitters or through direct stimulation of brain circuits [6]. The recognition that brain tumors actively exploit and integrate into existing neural circuits represents a groundbreaking shift in neuro-oncology. This perspective moves beyond a purely cell-autonomous view of cancer, acknowledging the profound influence of the nervous system on tumor progression [7]. This "cancer neuroscience" paradigm opens entirely new therapeutic avenues, suggesting that modulating these co-opted neural pathways could directly impede tumor growth and survival. Pharmacological neuromodulation, a key component of this evolving field, specifically employs drug molecules to target components of neural pathways that brain tumors co-opt for their proliferation and survival [7]. Unlike the broad cytotoxic effects of conventional chemotherapy, these agents aim for more precise interactions, thereby potentially reducing systemic side effects [8]. This approach signifies a move from merely eliminating cancer cells to a more nuanced strategy of "reprogramming" the tumor's interaction with its environment. Understanding and disrupting the tumor's "communication" with neurons is now considered as vital as direct cytotoxicity.

1.1 Molecular docking and its role in neuromodulatory drug discovery

Molecular docking plays a pivotal role in this specialized area of drug discovery. It is a computational technique that predicts the optimal binding orientation and affinity of a small molecule (ligand) to a protein receptor, offering atomistic insights into molecular recognition [9]. This method is a cost- and time-efficient procedure for drug design, particularly valuable for screening vast chemical libraries [10]. For central nervous system (CNS) cancers, molecular docking is indispensable for identifying novel therapeutic targets, evaluating potential drug candidates, and repurposing existing drugs by predicting their interactions with brain-specific targets involved in neuromodulation [11]. The increasing reliance on molecular docking and other *in silico* methods reflects a broader trend in drug discovery, driven by the complexity and prohibitive cost of traditional experimental methods. For brain cancer, where experimental models are particularly challenging and blood–

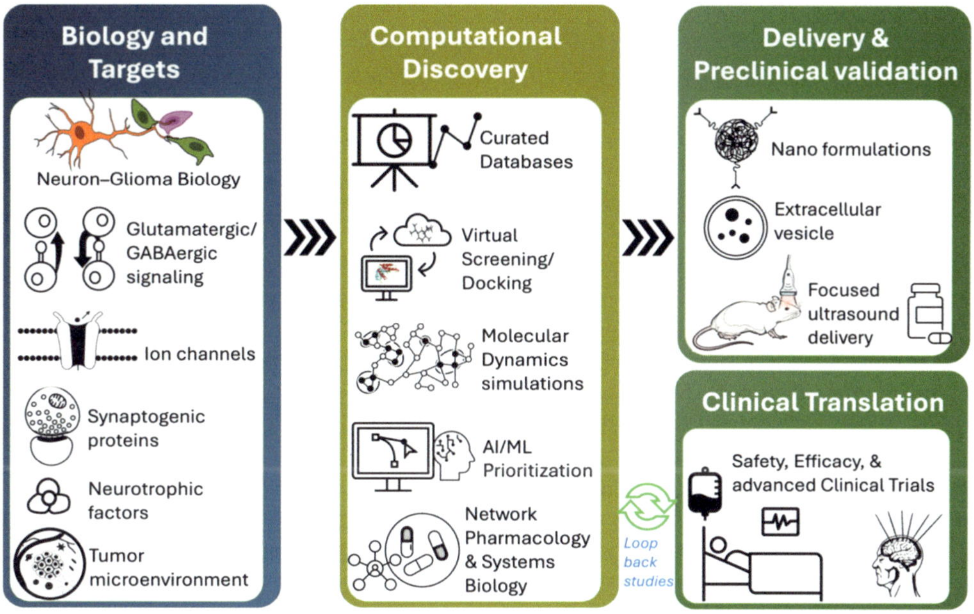

Figure 1.
Integrated translational pipeline from neuron-glioma biology to clinical translation. This schematic organizes the chapter's proposed workflow into four interconnected domains: (1) Biology and targets, which enumerates the neuron-glioma processes that define druggable vulnerabilities, that is, glutamatergic and GABAergic signaling, synaptogenic proteins, ion channels, neurotrophic factors, and the tumor microenvironment; (2) computational discovery, a linear discovery engine that progresses from curated databases, virtual screening/docking, molecular dynamics (MD) simulations, AI/ML prioritization, network pharmacology and systems biology to produce ranked ligands, predicted binding, and candidate target networks; (3) delivery and preclinical validation, which lists translational enablers and test platforms, that is, nano-formulations, extracellular-vesicle/exosome systems, and focused-ultrasound delivery, coupled with in vitro *and* in vivo *evaluation; and (4) clinical translation, which denotes safety, efficacy assessment and advanced clinical trials. An iterative "loop-back" arrow explicitly links clinical and preclinical readouts back into the computational discovery pipeline for model refinement and re-prioritization. Together, the figure visualizes a closed, multidisciplinary pathway for converting neuron-glioma mechanistic insight into prioritized, deliverable neuromodulatory therapies.*

brain barrier (BBB) penetration is a critical factor, computational tools become an essential first step for initial lead identification and optimization [12–15]. The content of this chapter is summarized in **Figure 1.**

Molecular docking is a cornerstone technique in structure-based drug discovery that predicts ligand binding modes and estimates relative affinities by sampling poses within a protein binding site and scoring intermolecular interactions. Docking enables rapid, cost-effective virtual screening of large libraries to prioritize compounds for downstream experimental testing and is particularly valuable when seeking repurposed neuromodulatory drugs with known CNS exposure profiles. Several widely used docking platforms (e.g., AutoDock/AutoDock Vina, Glide, MOE) support large-scale screens and have been applied previously to CNS targets and glioma-relevant proteins [9, 16–18]. Accurate docking for brain-relevant targets requires explicit attention to protein flexibility, membrane environment (for ion channels), and BBB constraints. Accordingly, best practice integrates flexible docking with MD to sample receptor conformations and validate pose stability and uses rescoring methods (MM-GBSA or other binding free energy calculations) and AI/ML-based filters to reduce false positives. Combining docking outputs with ADME and BBB-penetration predictions, and with network pharmacology mapping, substantially improves the translational relevance of *in silico* hits by prioritizing compounds that not only bind strongly *in silico* but are likely deliverable to tumor sites [12–15, 19–22].

2. Neuromodulation in brain cancer: Mechanisms and targets

Neuromodulation drugs for brain cancer represent an emerging therapeutic paradigm that targets not only the tumor cells themselves but also the complex neurobiological networks that support their growth, invasion, and resistance to treatment [23–25]. Recent research has revealed that brain tumors, particularly gliomas, do not exist in isolation; they actively engage in bidirectional communication with surrounding neurons, astrocytes, and immune cells, co-opting neurotransmitter systems such as glutamatergic and GABAergic signaling to promote proliferation and evade immune surveillance [26]. Neuromodulatory agents, ranging from ion channel modulators and synaptic transmission regulators to drugs that alter neuroinflammatory signaling, can strategically interfere with these tumor-neuron interactions. By disrupting aberrant neuronal activity that fuels tumor progression, these drugs can attenuate cancer-related neuroplasticity, reduce tumor invasiveness, and potentially enhance the efficacy of conventional treatments such as chemotherapy, radiotherapy, and immunotherapy. Importantly, neuromodulation offers the possibility of selective targeting, as many of these pathways are aberrantly activated in tumor-associated neural circuits but not in healthy brain tissue, reducing off-target toxicity. Beyond direct antitumor effects, neuromodulation drugs may also preserve or restore neurological function by counteracting tumor-induced network disruptions, improving patient quality of life. This dual benefit, which simultaneously weakens the tumor and protects brain function, positions neuromodulatory pharmacology as a compelling and potentially transformative addition to the neuro-oncology arsenal [26, 27]. Given the importance of neuromodulation drugs in cancer therapy, several strategies have been applied, including both non-pharmacological and pharmacological approaches.

2.1 Non-pharmacological neuromodulation approaches

Non-pharmacological neuromodulation techniques offer direct means to influence brain activity and, by extension, the tumor microenvironment (TME). These methods include transcranial magnetic/electrical stimulation (TMS/TES), vagus nerve stimulation (VNS), transcranial direct current stimulation (tDCS), deep brain stimulation (DBS), and focused ultrasound (FUS) [28–30]. (see **Figure 2**).

2.2 Pharmacological neuromodulation: Targeting neural pathways in gliomagenesis

The understanding that brain tumors actively integrate into and exploit neural circuits has led to a focused effort on pharmacological neuromodulation [24, 31, 32]. This involves designing or repurposing drugs that interfere with these tumor-promoting neural interactions. Brain tumor cells, particularly gliomas and metastatic cells, have been found to express functional receptors for various neurotransmitters

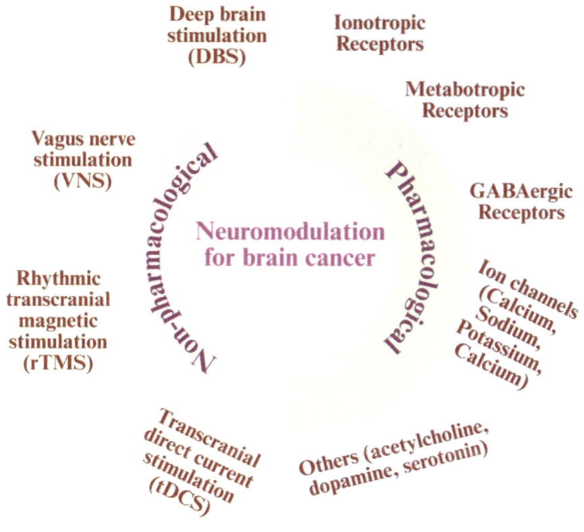

Figure 2.
Neuromodulation strategies for brain cancer: pharmacological and non-pharmacological approaches. The schematic illustrates two complementary modalities of neuromodulation currently being explored in the context of brain cancer therapy. Non-pharmacological neuromodulation encompasses device-based interventions including deep brain stimulation (DBS), vagus nerve stimulation (VNS), rhythmic transcranial magnetic stimulation (rTMS), and transcranial direct current stimulation (tDCS), which modulate neuronal excitability and network dynamics through electrical or magnetic fields. Pharmacological neuromodulation targets receptor and ion channel systems that regulate synaptic and extrasynaptic signaling in tumor–neuron interactions. These include ionotropic and metabotropic receptors, gabaergic receptors, voltage-gated ion channels (calcium, sodium, potassium), and neuromodulatory transmitters such as acetylcholine, dopamine, and serotonin. Together, these approaches highlight the dual therapeutic landscape where circuit-level modulation and molecular receptor targeting converge to reprogram aberrant neurophysiological signaling that drives tumor progression.

and to form "tumor synaptogenesis," integrating into existing neural networks to fuel their growth and survival [7].

2.2.1 Glutamatergic signaling (ionotropic and metabotropic receptors)

Glutamate, the most abundant excitatory neurotransmitter in the CNS, plays a critical role in brain tumor growth and proliferation [7]. This phenomenon, where glioma and metastatic cells exploit glutamatergic pathways, highlights a fundamental mechanism by which tumors hijack the brain's own communication systems for their benefit [7].

Ionotropic receptors (NMDA, AMPA, Kainate): AMPA receptors (AMPARs) are highly expressed on glioma cells. While their activation can sometimes trigger apoptosis, malignant glioma cells often release glutamate into the extracellular space, which then activates calcium-permeable AMPARs and oncogenic pathways like Akt, ultimately promoting glioma cell growth and migration [7]. NMDAR activation on GBM cells similarly promotes cell survival and migration, with upregulation observed in brain metastatic samples [7]. Antagonizing NMDARs can enhance radiosensitivity by impairing DNA double-strand break repair [7].

Metabotropic receptors (mGluRs): These G-protein-coupled receptors regulate intracellular signaling pathways. Group I mGluRs, such as mGluR3, are predominantly excitatory, and increased mGluR3 expression correlates with higher malignancy and fatality rates in GBM [7]. Inhibition of mGluR3 has been shown to sensitize GBM cells to Temozolomide, leading to reduced tumor growth [33]. Conversely, mGluR4 appears to suppress tumor growth and induce apoptosis [34]. The dual and sometimes contradictory roles of neurotransmitter receptors, such as AMPARs triggering apoptosis in some contexts while promoting growth in others, or the debated role of GABA, underscore the inherent complexity of tumor biology. This suggests that therapeutic targeting must be highly specific, carefully considering receptor subtypes, the cellular context (tumor vs. healthy cells), and the dynamic TME. A simplistic approach, such as broadly blocking all glutamate receptors, might not be effective or even safe.

2.2.2 Gabaergic signaling (GABAA and GABAB receptors)

GABA, the primary inhibitory neurotransmitter in the CNS, also exhibits complex and sometimes paradoxical effects on brain tumors [7].

GABAA receptors: Activated GABAA receptors have been observed to curb glioma cell proliferation and induce apoptosis in neuroblastoma cells [7]. However, endogenous GABAA receptor signaling can also sustain the quiescent state of a subset of tumor-initiating glioma stem-like cells, which are known to contribute to tumor recurrence and therapy resistance [7]. Furthermore, overexpression of glutamate decarboxylase 1 (GAD1), a key enzyme in GABA synthesis, has been linked to facilitated brain metastasis [7]. Paradoxically, tumor-promoting GABAergic neuron-to-glioma synapses, mediated by GABAA receptors, have been identified in diffuse midline gliomas (DMGs). These synaptic inputs depolarize DMG cells due to altered intracellular chloride concentrations, ultimately driving DMG cell proliferation *in vivo* [7].

GABAB receptors: Increased expression of GABAB receptors has been detected in various cancer cells. Agonists of GABAB receptors, such as baclofen, have shown promise in inhibiting tumor growth in rat models by suppressing the proliferation and migration of cancer cells [35].

GABA metabolism: Human breast cancer cells have been observed to proliferate and metastasize to the brain by catabolizing GABA into succinate, effectively utilizing GABA as an onco-metabolite. Inhibiting GABA synthesis (e.g., by targeting GAD67) or its utilization (e.g., by targeting GABA transaminase, ABAT) can attenuate cancer cell proliferation and migration [36].

Other neurotransmitters: Beyond glutamate and GABA, a wide array of neurotransmitters, including acetylcholine, dopamine, serotonin, norepinephrine, neuropeptides (e.g., substance P, neuropeptide Y), adenosine triphosphate (ATP), and gases like nitric oxide, have been implicated in various aspects of cancer biology, suggesting broader interactions between neurotransmitter systems and cancer progression. Notably, acetylcholinergic neurons have been shown to drive glioblastoma invasion [37].

2.3 Ion channels as therapeutic targets

Ion channels, which regulate the flux of ions such as sodium (Na^+), potassium (K^+), calcium (Ca^{2+}), and chloride (Cl^-) across cell membranes, are crucial for fundamental cellular physiology. They are increasingly recognized for their significant roles in tumor pathophysiology, influencing processes such as gene expression, cell migration, proliferation, glioma-related seizures, chemotherapy sensitivity, and tumor metabolism [38]. Their accessibility on the cell surface makes them particularly attractive "druggable" targets for neuromodulatory therapeutics [39].

Calcium channels (e.g., TRP channels, VGCCs): Higher expression of Ca^{2+} channels on glioma cells leads to increased Ca^{2+} influx, which, in turn, promotes Ca^{2+}-dependent proliferative signaling pathways [38]. Transient Receptor Potential Melastatin 7 (TRPM7) activates the JAK2/STAT3 and/or Notch signaling pathways, inducing cell proliferation and migration; its suppression significantly reduces these effects [38]. Midazolam, a benzodiazepine capable of crossing the BBB, has been shown to inhibit TRPM7 currents and calcium influx, leading to G0/G1 cell cycle arrest and decreased glioblastoma cell proliferation [38]. T-type Ca2 + channels (Cav3.2) are highly expressed in glioblastoma, and their pharmacological blockade inhibits tumor growth and enhances the effects of temozolomide [38].

Potassium channels (e.g., hEAG1, hERG, Kv1.5, Kv2.1): These channels are often overexpressed in GBM and are involved in signaling pathways that promote proliferation and inhibit apoptosis [38]. hERG blockers, for instance, have demonstrated therapeutic benefit in xenograft models and in glioblastoma patients [38]. KCND2, which encodes the voltage-gated K^+ channel KV4.2, has been implicated in enhancing neuronal excitability in epileptic GBM, suggesting it as a potential therapeutic target [39]. Furthermore, genes encoding ion transport proteins like GJB2 and SCN9A, part of the "ion permeome," show prominent expression in neoplastic cells and are associated with poor prognosis in glioblastoma; their knockdown impairs cell viability and tumor sphere formation in vitro and extends survival in xenograft models [40].

Chloride and sodium channels: Volume-regulated anion channel (VRAC) and Anoctamin-1 (ANO1), a Ca^{2+}-activated Cl^- channel, are involved in glioma cell proliferation [38]. Chloride Intracellular Channel-1 (CLIC1) is overexpressed in GBM, and its expression is associated with a worse prognosis, promoting proliferation [38]. Additionally, ASIC1 and Epithelial Sodium Channel (ENaC) play roles in cell migration and proliferation [38]. The widespread dysregulation of ion channels in brain cancer and their accessibility on the cell surface make them highly attractive "druggable" targets for neuromodulatory therapeutics. Molecular docking is instrumental in identifying and optimizing modulators for these complex membrane proteins.

2.4 Neuron-glioma synapses: A novel therapeutic frontier

Recent groundbreaking discoveries have confirmed the existence of direct synaptic connections between neurons and brain tumor cells, fundamentally altering the understanding of gliomas and brain metastases [24]. These neuron-glioma synapses are not merely incidental but actively drive tumor growth and invasion, particularly within the tumor infiltration zone [41]. Neurons contribute to tumor progression through both paracrine mechanisms, such as the secretion of neuroligin-3, and direct electrochemical mechanisms, notably through glutamatergic AMPA receptors [23]. Tumor cells also form extensive multicellular networks via tumor microtubules (TMs), which facilitate the distribution of small molecules like calcium, thereby promoting tumor cell survival [23]. Disturbingly, radiotherapy has been shown to enhance neuron-tumor connectivity by increasing neuronal activity, suggesting that this crosstalk contributes to therapeutic resistance [37].

Thrombospondin-1 and synaptogenesis: Thrombospondin-1 (TSP-1), a protein involved in neural circuit development and remodeling, has been identified as a key mediator of interactions between neurons and tumor cells that promote glioma growth [42]. This synaptogenic protein facilitates the functional integration of the tumor into the neural circuitry [42]. Gabapentin, a widely used anti-seizure and pain medication, has been shown to inhibit TSP-1, thereby disrupting neuronal synaptogenesis and neuronal activity-dependent glioblastoma proliferation [42]. Retrospective studies have linked gabapentin use to a significant survival benefit (four to six months longer median overall survival) in glioblastoma patients, a benefit potentially mediated through a reduction in serum TSP-1 levels [42]. The identification of specific molecular mediators like TSP-1 in neuron-glioma synaptogenesis provides a concrete, actionable target for pharmacological intervention. The observed survival benefit with gabapentin, linked to TSP-1 reduction, suggests that disrupting this neural integration is a viable anti-cancer strategy, even with existing, repurposed drugs.

Neurotrophic factors and their dual role: Neurotrophins, such as brain-derived neurotrophic factor (BDNF), are vital for the proper development and survival of neurons in the CNS. BDNF binds to its receptor, TrkB, initiating signaling cascades (e.g., via the PI3K/Akt pathway) that promote pro-migratory, anti-apoptotic, and pro-survival proteins in cells [43]. Historically, overexpression of BDNF/TrkB has been considered oncogenic, contributing to the progression of aggressive neuroblastoma and breast cancer stem cells [43]. However, recent evidence suggests a more complex, dual role for these factors. BDNF overexpression in the hypothalamus, for instance, has been shown to possess immune-augmenting properties, leading to an increased anti-tumor immune response and reducing resistance to chemotherapeutic agents *in vivo* [43]. This highlights that the effect of neurotrophic factors can be tissue-dependent and significantly influenced by the specific microenvironment [43]. The dual role of neurotrophic factors like BDNF presents a critical nuance in targeting neural pathways. What might be oncogenic in one context (e.g., direct tumor cell signaling) could be anti-tumorigenic in another (e.g., modulating immune response). This implies that therapeutic strategies must carefully consider the specific cellular context and the broader systemic effects, not just the direct impact on tumor cells.

3. Molecular docking and computational drug discovery in brain cancer neuromodulation

Molecular docking is a cornerstone of modern computer-aided drug design (CADD), offering a powerful computational framework to accelerate the discovery and optimization of drug candidates, particularly for complex diseases like brain cancer.

3.1 Principles and applications of molecular docking

Molecular docking is a computational procedure designed to predict the optimal, lowest-energy binding conformations of a small molecule (ligand) within the binding site of a biological macromolecule, typically a protein receptor, and to estimate the binding affinity between them [16]. This process involves quantifying various intermolecular interactions, including shape and electrostatic complementarities, van der Waals forces, and hydrogen bond formation [17]. The sum of these interactions is approximated by a docking score, which serves as an indicator of the potential for binding [17].

The applications of molecular docking are diverse and strategically important in drug discovery:

Predicting ligand–target interactions and binding affinities: Molecular docking provides atomistic insights into precisely how drug molecules interact with their intended biological targets [18]. This detailed understanding is crucial for rational drug design, allowing researchers to predict which compounds are most likely to bind effectively and with high affinity.

Virtual screening for novel compounds: One of the most impactful applications is the rapid in silico screening of vast chemical libraries to identify potential drug candidates [11]. This significantly reduces the time and financial investment required for traditional high-throughput experimental screening, enabling a much more efficient initial filtering of compounds.

Repurposing existing drugs: Molecular docking is instrumental in identifying new therapeutic uses for approved drugs by predicting their interactions with novel targets [44]. This approach is particularly valuable for brain cancer, given the high failure rate of new CNS drug development due to BBB penetration issues. Repurposing existing drugs with known safety profiles and some BBB permeability, such as gabapentin and perampanel, can significantly accelerate clinical translation [42]. The ability to leverage the established safety and pharmacokinetic profiles of existing medications offers a de-risked and faster pathway to new brain cancer treatments.

3.2 Key molecular docking software and algorithms

The field of molecular docking has seen the development of over 60 different tools and programs for both academic and commercial use over the past two decades [17]. These tools employ various strategies for ligand placement, including incremental construction, shape-based algorithms, genetic algorithms, systematic search techniques, and Monte Carlo simulations [17].

Commonly utilized software in the field includes AutoDock (and its optimized version, AutoDock Vina), FlexX, Surflex, GOLD, ICM, Glide, MOE-Dock, and LeDock [18]. For instance, AutoDock 4.2 has been extensively used in studies investigating the interactions of Temozolomide with glioblastoma-related proteins [45, 46]. MOE software has been employed to confirm the high binding affinity of compounds like TIZ to targets such as CDK1 in glioblastoma [47]. PyRx is another common tool used for docking phytochemicals [48].

A critical consideration for CNS targets is accounting for protein dynamics and flexibility. While many docking programs simplify the process by treating the protein receptor as rigid or allowing only limited flexibility to residues near the active site, this can lead to inaccuracies [18]. Proteins are dynamic entities, fluctuating through various conformations, and ligands often interact with specific, transient structural states [19]. This "rigid receptor" assumption is a significant limitation, especially for dynamic CNS targets like ion channels and G-protein coupled receptors (GPCRs), which are known for their substantial conformational changes upon ligand binding [49]. Flexible docking methods, which consider multiple possible conformations of both the ligand and the receptor, are therefore more accurate for predicting binding modes and affinities, though they significantly increase computational complexity. The necessity for flexible docking and MD simulations underscores that simplistic models are often insufficient for accurate predictions in the complex and dynamic brain environment.

3.3 Integration with advanced computational methods

To enhance the understanding of molecular docking's role in neuromodulatory drug discovery for brain cancer, its integration with complementary computational methods is essential for addressing limitations like protein flexibility and binding accuracy. As illustrated in **Figure 3**, molecular docking acts as a central node, synergizing with artificial intelligence (AI) and statistical methods for pre-docking screening (e.g., protein conformation selection and scoring function improvement) and post-docking pose rescoring; binding free energy methods for post-docking pose rescoring; MD simulations for pre-docking identification of representative conformations and post-docking assessments of pose refinement and ligand-target complex stability; and ligand-based approaches for pre-docking protein conformation selection and post-docking pose selection/rescoring. This multifaceted

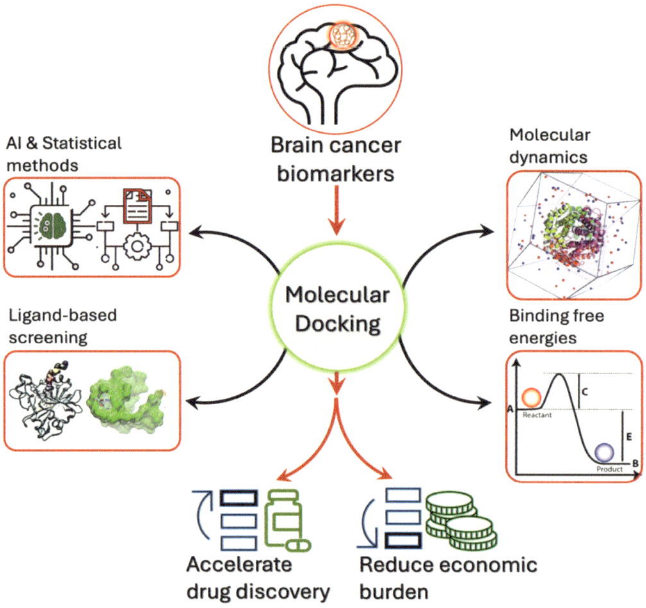

Figure 3.
Molecular docking's role in neuromodulatory drug discovery for brain cancer and its integration with complementary computational methods. Molecular docking acts as a central node, synergizing with artificial intelligence (AI) and statistical methods for pre-docking screening (e.g., Protein conformation selection and scoring function improvement) and post-docking pose rescoring; binding free energy methods for post-docking pose rescoring; molecular dynamics (MD) simulations for pre-docking identification of representative conformations and post-docking assessments of pose refinement and ligand–target complex stability; and ligand-based approaches for pre-docking protein conformation selection and post-docking pose selection/rescoring. This multifaceted integration accelerates virtual screening and optimization of neuromodulatory agents targeting glioma–neuron interactions, such as glutamate or GABA receptors, while improving predictions for blood-brain barrier-penetrant compounds, which would ultimately result in accelerated drug discovery and reducing the overall cost.

integration accelerates virtual screening and optimization of neuromodulatory agents targeting glioma-neuron interactions, such as glutamate or GABA receptors, while improving predictions for BBB-penetrant compounds [20–22].

AI/Machine learning in drug discovery: AI and Machine Learning (ML) are rapidly gaining prominence for their ability to process vast chemical datasets, identify complex patterns, and predict compounds with high therapeutic potential [44]. These technologies can derive predictive characteristics from diverse data sources, including imaging and genomic sequences, and model tumor development dynamics and drug-tumor interactions [50]. AI-powered platforms, such as DrugBank, are already leveraging AI to uncover novel connections and insights in drug–target interactions, streamlining the discovery process [51].

MD simulations: MD simulations are theoretical tools that explore the configurations and dynamic behaviors of molecules over time, providing atomic-level insights into drug mechanisms of action [11]. They are essential for evaluating the flexibility and stability of protein-ligand complexes, thereby overcoming the inherent rigid-receptor limitation of basic docking approaches [17]. MD simulations can reveal how a drug interacts with a dynamic target, providing a more realistic picture of binding.

Network pharmacology and systems biology approaches: These methodologies analyze molecular and genetic mechanisms at a systems level, identifying core driving genes and abnormal regulation pathways within complex biological networks [52]. They are crucial for understanding intricate protein–protein interactions and broader signaling pathways, providing a holistic view of disease mechanisms and drug effects [52]. The convergence of molecular docking with MD simulations, AI/ML, and network pharmacology is critical for tackling the immense complexity of brain tumor biology, particularly the dynamic neuron–glioma interactions and the heterogeneous TME. This integrated approach allows researchers to move beyond simple ligand–receptor binding to understand complex biological systems and the broader impact of drugs.

3.4 Relevant drug–target interaction databases

High-quality, integrated databases form the backbone of modern computational drug discovery. Their ability to link drugs to targets, diseases, and even genetic features of cancer cells is fundamental for identifying neuromodulatory compounds and validating *in silico* predictions.

DrugBank: DrugBank is a comprehensive online database that provides reliable and up-to-date information on drugs and their targets. It offers real-time insights, increasingly powered by AI, and integrates essential external datasets. This resource is widely utilized by scientists, researchers, and biotech teams to discover hidden drug–target interactions and explore extensive datasets. DrugBank lists "Brain (Nervous System) Cancers" as a relevant indication, allowing for targeted searches [53].

Cancer therapeutics response portal: The cancer therapeutics response portal (CTRP) is a valuable resource that links genetic, lineage, and other cellular features of cancer cell lines to small-molecule sensitivity [54]. It contains an "Informer Set" of 481 compounds rigorously tested across 860 cancer cell lines, providing annotations for small molecules by protein target and for cell lines by mutation and lineage [54]. CTRP serves as a dynamic resource for generating insights into small-molecule action mechanisms and formulating novel therapeutic hypotheses. The existence and quality of these integrated databases are crucial enablers for the entire computational drug discovery pipeline, allowing for data-driven hypothesis generation and validation, and addressing the critical need for robust datasets in drug discovery [19].

4. Promising neuromodulatory drug molecules and molecular docking insights

The application of molecular docking and related computational methods has facilitated the identification and characterization of several promising neuromodulatory drug molecules for brain cancer therapeutics. These agents often target neurotransmitter receptors, ion channels, or other neural-related proteins that are aberrantly involved in tumor progression.

4.1 Neurotransmitter-based modulators

Brain tumors can hijack neurotransmitter systems to promote their growth and survival, making neurotransmitter receptors attractive therapeutic targets. Computational approaches, including molecular docking, have helped identify several neurotransmitter-targeted modulators with potential anti-tumor effects. Below, we examine key examples targeting glutamate and GABA pathways:

4.1.1 Glutamate receptor modulators

Perampanel (AMPA antagonist): Perampanel (PER) is a highly selective, non-competitive antagonist of glutamate α-amino-3-hydroxy-5-methyl-4-isoxazolepropionic acid (AMPA) receptors [55]. While primarily approved for treating seizures, including brain tumor-related epilepsy (BTRE), PER has demonstrated promising antitumor effects in *in vitro* studies. It inhibits glioblastoma cell proliferation, attenuates glucose uptake, and reduces extracellular glutamate levels [55]. PER also exhibits good BBB penetration, a crucial property for CNS drugs [55]. However, despite encouraging *in vitro* results, its anti-tumor effects have not been consistently confirmed in preclinical studies, and clinical data, specifically on its impact on tumor progression, remain scarce. The contrast between *in vitro* efficacy and limited *in vivo* or clinical success for many glutamate antagonists strongly underscores the persistent challenges of BBB penetration. This highlights that successful neuromodulatory drugs for brain cancer must either possess inherent BBB permeability (like perampanel or gabapentin) or be delivered via BBB-disrupting technologies (like FUS) [55].

Gabapentin (α2δ-1/TSP-1 inhibition): Gabapentin, a widely used anti-seizure and neuropathic pain medication, exerts its effects by decreasing the activity of the α2δ-1 protein [56]. Crucially, α2δ-1 has been identified as the neuronal receptor for TSP-1 [56]. TSP-1 is a synaptogenic protein that actively promotes the functional integration of tumors into neural circuitry, thereby fueling glioma growth [42]. By antagonizing TSP-1 binding to α2δ-1, gabapentin effectively inhibits excitatory synapse formation [56]. Retrospective studies have linked gabapentin use to a significant survival benefit (an average of four to six months longer median overall survival) in glioblastoma patients, a benefit potentially mediated through a reduction in serum TSP-1 levels [46]. This promising finding warrants further prospective clinical trials to validate its therapeutic potential. The identification of specific molecular mediators like TSP-1 in neuron-glioma synaptogenesis provides a concrete, actionable target for pharmacological intervention. The observed survival benefit with gabapentin, linked to TSP-1 reduction, suggests that disrupting this neural integration is a viable anti-cancer strategy, even with existing, repurposed drugs.

Other glutamate antagonists: Other glutamate antagonists, such as dizocilpine (an NMDA antagonist) and GYKI52466 (an AMPA antagonist), have demonstrated antiproliferative effects in various human tumor cell lines, including astrocytoma. These effects are attributed to decreased cell division and increased cell death [57]. However, similar to other CNS-targeted agents, their poor efficacy in penetrating the BBB has significantly limited their clinical application for brain tumors [57]. Fluoxetine, a selective serotonin reuptake inhibitor (SSRI), also exhibited calcium-dependent cytotoxicity in glioma cells, requiring transmembrane calcium influx [58]. Talampanel, another AMPA receptor antagonist, showed mixed results in clinical trials, further emphasizing the challenges of translating *in vitro* efficacy to consistent clinical benefit [55].

4.1.2 GABA receptor modulators

GABAA and GABAB receptor agonists/antagonists: As discussed previously, activation of GABAA receptors has been shown to curb glioma cell proliferation and induce apoptosis in neuroblastoma cells [7]. Conversely, GABAB receptor agonists like baclofen have demonstrated promise in inhibiting tumor growth by suppressing proliferation and migration [59]. Molecular docking studies have been performed for GABAA receptor modulators, such as pyrazoloquinolinones (PQs), providing insights into their binding modes and interactions with specific receptor subunits [60].

Valproic acid (VPA): Valproic acid (VPA) is an antiepileptic drug frequently used in glioblastoma patients for seizure control [82]. Beyond its anticonvulsive properties, VPA is also a histone deacetylase inhibitor (HDACi), primarily targeting Class I and IIa HDACs [61]. Molecular docking and dynamics studies have elucidated VPA's binding to the catalytic site of various HDACs, providing structural and energetic insights into its inhibitory mechanism [62]. VPA has been shown to sensitize glioma cells to temozolomide by upregulating solute carrier (SLC) transporters and promoting the reexpression of tumor suppressor genes [61]. Furthermore, VPA reduces cell proliferation, induces cell cycle arrest, differentiation, and apoptosis, partly by increasing the production of reactive oxygen species (ROS) [61]. Valproic acid exemplifies a repurposed drug with dual mechanisms relevant to brain cancer: direct neuromodulation (anti-epileptic effects) and epigenetic modulation (HDAC inhibition). This highlights that drugs may exert anti-cancer effects through multiple, interconnected pathways, which can be elucidated and optimized through molecular docking and systems-level analysis.

4.2 Ion channel modulators

Ion channels are a significant class of drug targets, with approximately 10%–20% of small-molecule drugs targeting them [63]. Their widespread dysregulation in brain cancer and their accessibility on the cell surface make them highly attractive targets for neuromodulatory therapeutics.

4.2.1 Specific examples and mechanisms

TRPM7 inhibitors: TRPM7 is a calcium-permeable cation channel linked to glioma cell survival and proliferation. Inhibiting TRPM7 can stunt tumor growth. For example, the anesthetic midazolam was found to acutely block TRPM7 channel activity in glioblastoma cells. Even a few seconds of midazolam exposure suppressed TRPM7 currents and downstream Ca^{2+} influx, while longer exposures reduced cell proliferation. This suggests TRPM7 channel blockade (by midazolam or other agents) can induce calcium starvation and growth arrest in glioma cells. Carvacrol and other TRPM7 modulators have similarly shown anti-glioma effects in preclinical models. Targeting TRPM7, which also interacts with signaling pathways (Notch, PI3K/mTOR), represents a potentially novel anti-glioma strategy that is currently under investigation.

T-type Ca^{2+} channel blockers: T-type calcium channels (Cav3.x) help regulate cell cycle transitions. Mibefradil, a T-type Ca^{2+} channel blocker formerly used for hypertension, was repurposed for glioma based on preclinical efficacy. Mibefradil inhibited glioblastoma growth and notably sensitized tumors to TMZ in both cell culture as well as rodent models. In glioma stem cell-derived xenografts, oral mibefradil slowed tumor progression and extended survival when combined with TMZ. Furthermore, a Phase I/II trial (NOA-08/Persurge) tested sequential mibefradil + TMZ in recurrent GBM, demonstrating safety and some pharmacodynamic effects, though further trials are still needed to establish efficacy. T-channel blockers prevent the Ca^{2+} influx needed for G1/S cell-cycle progression, thereby arresting tumor cells. This mechanism, coupled with synergy with chemotherapy, makes mibefradil and newer Cav3 inhibitors potential candidates.

hERG potassium channel blockers: The hERG (Kv11.1) potassium channel, better known for its role in cardiac repolarization, is also expressed in gliomas. High hERG expression in GBM correlates with poorer prognosis. Some clinically approved drugs incidentally block hERG channels (e.g., certain anti-arrhythmics, antipsychotics). Retrospective analyses suggest that GBM patients on hERG-blocking drugs had improved survival. Specifically, patient populations whose tumors highly expressed hERG and those receiving one or more hERG blockers lived significantly longer. Further, hERG channel blockers showed anti-glioma effects in xenograft studies, potentially by inducing apoptosis or by sensitizing cells to therapy. Despite this, caution is needed owing to the risk of cardiac side effects (QT prolongation) with hERG blockade, and as such, efforts have now focused on non-torsadogenic hERG inhibitors or localized delivery to reduce the systemic risks. Nonetheless, the data identify hERG as a functional player in glioma cell viability and a possible therapeutic target when carefully managed.

Gap junctions and sodium channels: Beyond classical ion channels, glioma cells also exploit channels involved in cell–cell communication and excitability. A recent large-scale pan-cancer analysis identified GJB2 (connexin 26, a gap junction protein) and SCN9A (Nav1.7, a voltage-gated sodium channel) as key pro-tumor ion transport genes in GBM. It was found that knocking down GJB2 or SCN9A in patient-derived GBM cells had a drastic effect on cell viability, and it impaired the formation of tumor spheres (a measure of self-renewal). These results suggest that electrical coupling (via gap junctions) and sodium current activity support glioma cell survival and stemness. While no specific drugs targeting connexin 26 or Nav1 [5] in GBM are in use yet, these hold promise as emerging targets. Strategies like gap junction inhibitors or selective sodium channel blockers (already used in epilepsy and pain) could be explored as adjuncts to disconnect tumor cells from supportive neural/glial networks.

4.3 Other targeted neuromodulatory agents

Beyond direct neurotransmitter and ion channel modulators, other agents and delivery strategies hold neuromodulatory implications.

FUS-gated nanoparticle delivery: Advances in biophysics now allow us to temporarily breach the BBB or deliver drugs *in situ* using noninvasive FUS. FUS, when combined with microbubbles or smart drug carriers, can localize therapy to the tumor region with minimal systemic side effects. For example, FUS has been used to open the BBB transiently, allowing intravenous drugs to perfuse into brain tumors at high concentrations. In addition, researchers have developed nanoparticles that respond to ultrasound triggers. A striking case is the use of a nanoemulsion carrying the anesthetic propofol. In this, when ultrasound is applied to the target region, the already-present nanoemulsion releases propofol on demand. This technique enabled a noninvasive, focal neuromodulation paradigm where propofol (a GABA_A agonist) was delivered into specific brain areas, causing local neural inhibition without systemic anesthesia. Such FUS-gated therapies have immediate applications, for instance, selectively calming hyperexcitable tumor-infiltrated networks or improving drug delivery to invasive tumor margins. The approach has already entered clinical translation for gliomas, where FUS paired with microbubbles has been used to enhance chemotherapy delivery in GBM patients (in clinical trials like NCT03744026). Additionally, efforts are underway to integrate FUS with drug-loaded liposomes/nanoparticles for precision therapy. While not a "drug molecule" per se, FUS is a powerful enabling technology that can amplify the impact of neuromodulatory drugs by overcoming the BBB, which has proven to be a recurring hurdle in neuro-oncology.

Traditional targeted therapies with neuromodulatory implications: While not strictly "neuromodulatory" in their primary mechanism of action, some traditional targeted therapies for brain cancer interact with pathways that can indirectly influence the tumor's microenvironment or its interaction with neural elements. For instance, Bevacizumab targets vascular endothelial growth factors (VEGF), inhibiting the formation of new blood vessels that supply tumors. Everolimus blocks mTOR, a protein crucial for cell growth and division. Temozolomide, a DNA-alkylating agent, has been shown via molecular docking studies to interact with several secretory proteins (GDF1, SLIT1, NPTX1, CREG2, SERPINI) implicated in gliomagenesis and radio resistance [45]. The epidermal growth factor receptor (EGFR) is another significant target in GBM, with mutations like EGFRvIII leading to constitutive activation of mitogenic signaling pathways [64]. The overlap between traditional targeted therapies and neuromodulation, particularly in the context of the TME and cell signaling, suggests that a holistic, interdisciplinary approach is needed. Drugs initially designed for one purpose might have unexpected neuromodulatory effects that can be leveraged for therapeutic benefit.

Table 1 summarizes key neuromodulatory drug molecules, their molecular targets, and relevant insights from docking and mechanistic studies in the context of brain cancer.

Drug name (type)	Primary target(s)	Mechanism of action (neuromodulatory and anti-cancer)	Key findings/molecular docking insights	Ref (s)
Perampanel (AMPA antagonist)	AMPA-type glutamate receptors	Non-competitive AMPA receptor blocker; reduces excitatory transmission and glutamate-driven proliferation.	BBB penetrant. Inhibits glioma proliferation in vitro but has limited in vivo efficacy. Used clinically for BTRE seizure control.	[55]
Gabapentin (α2δ-1 ligand)	α2δ-1 calcium channel subunit (Thrombospondin-1 receptor)	Blocks TSP-1-induced excitatory synaptogenesis by binding the α2δ-1 subunit.	Inhibits neuron-glioma synapses. Retrospective study: + 4–6 months survival in GBM patients with lower serum TSP-1.	[56]
Dizocilpine (MK-801) GYKI-5,2466 (+ Talampanel)	NMDA receptor (MK-801); AMPA receptors (GYKI & talampanel)	Ionotropic glutamate receptor antagonists block Ca^{2+} currents and excitotoxic signaling.	Broad antiproliferative effects in vitro, but poor BBB penetration. Talampanel showed a modest survival benefit in Phase II.	[65]
Fluoxetine (SSRI antidepressant)	AMPA receptors (on tumor cells); Serotonin transporters (in the CNS)	Allosteric AMPAR modulator causing excessive Ca^{2+} influx and apoptosis in glioma cells.	Selective glioma cell killing via Ca^{2+}-mediated apoptosis. Oral treatment suppressed GBM growth, comparable to TMZ.	[58]
Baclofen (GABA_B agonist)	GABA_B receptors (GPCR, metabotropic GABA)	Activates inhibitory GABA_B signaling; reduces cAMP and pro-growth pathways.	Suppressed tumor growth in rodent models via GABA_B-mediated inhibition of CREB/ERK signaling.	[66]
Pyrazoloquinolinones (PQs) (GABA_A receptor modulators)	GABA_A receptor (extracellular benzodiazepine site at α1 +/β3– interface)	Allosteric GABA_A potentiators enhancing GABAergic inhibition.	Molecular docking validated α1–β3 interface binding. Tool compounds for modulating neuronal–tumor interactions.	[60]
Valproic Acid (VPA) (antiepileptic and HDAC inhibitor)	Neuronal target: Increases GABA levels. Tumor target: Class I and IIa HDAC enzymes.	Enhances GABA neurotransmission and inhibits HDACs for epigenetic tumor suppression.	Sensitizes GBM to TMZ via SLC upregulation and DNA-repair downregulation. Induces cell-cycle arrest and differentiation.	[67]
Midazolam (Benzodiazepine anesthetic)	TRPM7 channels (in glioma cells); GABA_A receptors (in neurons – sedation effect).	GABA_A modulator (sedation) and TRPM7 channel blocker (anti-proliferative).	Suppresses TRPM7 currents and glioma proliferation. Lipophilic, BBB-crossing, but limited by CNS depression.	[68]
Mibefradil (T-type Ca^{2+} channel blocker)	Cav3 (T-type) calcium channels (+ some effect on L-type Ca channels)	Blocks T-type Ca^{2+} channels; prevents Ca^{2+} influx needed for cell cycle progression.	Induced tumor cell cycle arrest and enhanced TMZ efficacy in preclinical models. A Phase I/II trial showed tolerability.	[69]

Drug name (type)	Primary target(s)	Mechanism of action (neuromodulatory and anti-cancer)	Key findings/molecular docking insights	Ref (s)
hERG channel blockers (e.g., astemizole, etc.)	hERG (Kv11.1) potassium channels (often misexpressed in tumors)	Block K^+ currents are important for tumor electrophysiology and cell cycle regulation.	High hERG expression correlates with worse GBM outcomes. Patients on hERG blockers showed improved survival.	[70]
Thioridazine (Antipsychotic, EAG2 blocker)	EAG2 (Kv10.2) potassium channels; D2 dopamine receptors (original CNS target).	EAG2 K^+ channel antagonist impairing cell volume regulation during proliferation.	Significantly reduced medulloblastoma growth and eliminated spinal metastases in preclinical models.	[71]
Propofol (FUS-nanoparticle) (Focused Ultrasound delivery)	GABA_A receptors (β-subunit binding site on GABA_A – propofol's anesthetic target)	GABA_A-positive modulator delivered via FUS-triggered nanoparticles for targeted neuromodulation.	FUS-triggered propofol release achieved localized neuromodulation with negligible systemic exposure.	[72]
Bevacizumab (Anti-VEGF antibody)	VEGF-A (vascular endothelial growth factor) in tumor vasculature	Neutralizes VEGF to inhibit angiogenesis; reduces tumor edema and intracranial pressure.	Rapidly improves GBM symptoms through reduced edema. Prolongs progression-free survival but not overall survival.	[73]
Everolimus (mTOR inhibitor)	mTOR kinase (component of the mTORC1 complex)	Allosterically inhibits mTORC1; downregulates the PI3K–AKT–mTOR growth pathway.	Cytostatic effects in high-grade gliomas. Tumor shrinkage and seizure reduction in TSC-associated low-grade gliomas.	[74]
Temozolomide (TMZ) (DNA alkylator)	Genomic DNA in tumor cells (plus potential secreted protein targets).	DNA methylation at O6-guanine, triggering mismatch repair and apoptosis.	Standard GBM chemotherapy. Potential off-target binding to neuronal proteins (GDF1, SLIT1), affecting the tumor microenvironment.	[75]

Table 1.
Key neuromodulatory drug molecules, their molecular targets, and relevant docking/mechanism insights in brain cancer.

5. Challenges and limitations in molecular docking for brain cancer neuromodulation

Despite the advances in molecular docking for developing neuromodulatory drugs for brain cancer, several significant barriers hinder their translation into clinical practice.

BBB penetration: The BBB restricts most drugs from entering the CNS, allowing only small, lipophilic molecules (typically under 400–500 Da) to pass. Efflux pumps, like P-glycoprotein, further limit access to the CNS, and while chemical modifications can enhance permeability, they often compromise the drug's efficacy. Possible solutions include using FUS to disrupt the BBB and employing integrated models to evaluate permeability at early stages.

Tumor heterogeneity and microenvironment: Glioblastomas exhibit extreme genetic and phenotypic diversity, along with complex TMEs that include immune, vascular, and stromal components. Phenomena such as immune evasion – exemplified by the upregulation of PD-1/PD–L1 – and changes driven by tumor recurrence render single-target docking methods inadequate for capturing the dynamics of multiple cell types.

Docking limitations: Assumptions regarding protein rigidity can overlook induced-fit effects, and scoring functions often lack universal accuracy, leading to false positives. The absence of standardized datasets makes reproducibility challenging, and large-scale screening requires substantial computational resources.

Translational gaps: Beyond technical limitations, issues related to the BBB, tumor heterogeneity, and the complexities of trial design slow down clinical progress. Docking approaches must integrate biological insights, rigorous in vitro and in vivo validation, and well-designed clinical trials to ensure therapeutic success.

The computational strategy positions molecular docking at the core of neuromodulatory drug discovery for brain cancer. Docking provides the primary triage for large chemical libraries, generating ranked ligand lists and predicted binding poses that are subsequently validated by MD, rescoring, and ADME/BBB filters. Emphasizing rigorous docking workflows, such as careful receptor preparation, explicit treatment of flexibility, pose rescoring, and cross-platform comparison, will reduce false positives and streamline experimental prioritization.

Molecular docking is not merely an exploratory screen but the central decision step of the pipeline: it rapidly narrows chemical space to a manageable set of candidates for downstream experimental validation. To maximize translational yield, docking must be executed as part of an integrated workflow, such as ensemble/flexible docking to account for receptor dynamics, rescoring with physics-based methods (MM-GBSA or FEP where feasible), MD validation of binding pose stability, and ADME/BBB filtering to prioritize deliverable compounds. We therefore recommend reporting both docking and rescoring metrics, presenting representative pose visualizations, and benchmarking the pipeline against reference ligands (e.g., perampanel for AMPAR, gabapentin for $\alpha 2\delta$-1) to demonstrate predictive value prior to experimental testing.

6. Future directions and recommendations

Developing neuromodulatory drugs for brain cancer requires a multi-pronged strategy that integrates advanced computational tools, innovative delivery systems, personalized medicine, and robust translational research. AI-driven drug discovery, deep learning, and integrated multi-scale modeling are essential for navigating the complexity of tumor biology, improving docking accuracy, and efficiently repurposing compounds. Overcoming the BBB requires innovative delivery methods, including FUS, targeted nanoparticles, radioconjugates, and exosome-based systems to achieve accurate drug localization. Personalized approaches, guided by multi-omics data and predictive biomarkers, will enable therapies tailored to individual tumor profiles. Connecting findings from computer simulations with thorough laboratory, animal, and clinical studies is essential for confirming their effectiveness and safety. Ultimately, progress will depend on collaboration across various fields and the sharing of open-access data to accelerate innovation and its application from the lab to patient care.

7. Conclusion

The landscape of brain cancer therapeutics is undergoing a transformative shift, driven by the burgeoning field of cancer neuroscience and the recognition that brain tumors actively exploit neural pathways for their growth and survival. Neuromodulation, both pharmacological and non-pharmacological, offers a compelling new paradigm for intervention, moving beyond traditional cytotoxic approaches to disrupt the intricate interplay between tumor cells and the neural microenvironment. Molecular docking stands as an indispensable computational tool in this endeavor, enabling the prediction of drug–target interactions, the virtual screening of vast chemical libraries, and the repurposing of existing drugs with known safety profiles. Promising candidates, such as perampanel and gabapentin, demonstrate the potential of targeting neurotransmitter receptors and synaptogenic proteins like TSP-1, leading to observed anti-proliferative effects and, in the case of gabapentin, a survival benefit in glioblastoma patients. Similarly, the multi-mechanistic action of valproic Acid, combining neuromodulatory and epigenetic effects, underscores the value of identifying compounds with pleiotropic anti-cancer activities. Ion channels, widely dysregulated in brain tumors and accessible on the cell surface, also represent highly attractive and "druggable" targets for novel neuromodulatory therapeutics. Despite these advancements, significant hurdles persist. The BBB remains a formidable physiological barrier, limiting the brain penetration of many promising drug candidates. The profound intra- and inter-tumoral heterogeneity of brain cancers, coupled with the complex and dynamic TME, poses challenges for single-target therapies. Furthermore, inherent limitations in molecular docking methodologies, particularly regarding accurate protein dynamics, scoring functions, and the need for standardized validation, necessitate continuous refinement of computational models. To overcome these challenges and translate in silico predictions into meaningful clinical outcomes, a concerted, multidisciplinary effort is required. This includes advancing computational models through AI and multi-scale simulations, developing innovative drug delivery strategies like FUS-enhanced nanoparticle delivery and exosome-mediated transport, and embracing personalized medicine approaches guided by comprehensive biomarker identification. Ultimately, rigorous experimental validation in vitro and in vivo, followed by well-designed prospective clinical trials, remains paramount to confirm the safety and efficacy of neuromodulatory drug molecules. Collaborative research and open data-sharing initiatives will be critical to accelerating discovery and ensuring that these innovative therapeutic strategies reach patients in need, offering renewed hope in the fight against brain cancer.

Conflict of Interest

The authors declare no conflict of interest.

Author details

Monochura Saha[1,2]*, Shubham Yadav[1], Habiba Zeb[3] and Ishaq N. Khan[1,3]

1 Media Lab, Massachusetts Institute of Technology, Cambridge, USA

2 Birla Institute of Technology, Mesra, Ranchi, Jharkhand, India

3 Khyber Medical Univeristy, Peshawar, Pakistan

*Address all correspondence to: msaha@media.mit.edu

References

[1] Rabab'h O, et al. Twenty years after glioblastoma multiforme diagnosis: A case of long-term survival. Cureus. 2021;**13**(6):e16061

[2] Abou Jaoude D, et al. Glioblastoma and increased survival with longer chemotherapy duration. Kansas Journal of Medicine 2019;**12**:65

[3] Chiariello M, et al. Overcoming challenges in glioblastoma treatment: Targeting infiltrating cancer cells and harnessing the tumor microenvironment. Frontiers in Cellular Neuroscience. 2023;**17**:1327621

[4] Angom RS, Nakka NMR, Bhattacharya S. Advances in glioblastoma therapy: An update on current approaches. Brain Sciences. 2023;**13**(11):1536

[5] Boccuni L, et al. Neuromodulation-induced prehabilitation to leverage neuroplasticity before brain tumor surgery: A single-cohort feasibility trial protocol. Frontiers in Neurology. 2023;**14**:1243857

[6] Krames ES, et al. What is neuromodulation? Neuromodulation. Elsevier; 2009. p. 3–8

[7] Mondal J, Huse JT. Neurotransmitter power plays: The synaptic communication nexus shaping brain cancer. Acta Neuropathologica Communications. 2025;**13**(1):85

[8] Sarker S, Newland B. Repurposing neurological drugs for brain cancer therapeutics: A systematic approach to identify charged molecules for affinity-based local drug delivery systems. International Journal of Pharmaceutics. 2025;**682**:125935

[9] Dias R, de Azevedo J, Walter F. Molecular docking algorithms. Current Drug Targets. 2008;**9**(12):1040–1047

[10] Alikhani R, et al. Response Surface Study on Molecular Docking Simulations of Citalopram and Donepezil as Potent CNS Drugs. Iranian Journal of Pharmaceutical Research: IJPR 2021;**20**:560

[11] Salman MM, et al. Advances in applying computer-aided drug design for neurodegenerative diseases. International Journal of Molecular Sciences 2021;**22**:4688

[12] Alves PA, et al. Computational modeling of pharmaceuticals with an emphasis on crossing the blood–brain barrier. Pharmaceuticals 2025;**18**:217

[13] Spielvogel CP, et al. Enhancing Blood–Brain Barrier Penetration Prediction by Machine Learning-Based Integration of Novel and Existing, In Silico and Experimental Molecular Parameters from a Standardized Database. Journal of Chemical Information and Modeling 2025;**65**: 2773–2784

[14] Gupta M, Feng J, Bhisetti G. Experimental and computational methods to assess central nervous system penetration of small molecules. Molecules. 2024;**29**(6):1264

[15] Huang ET, et al. Predicting blood–brain barrier permeability of molecules with a large language model and machine learning. Scientific Reports. 2024;**14**(1):15844

[16] Brooijmans N, Kuntz ID. Molecular recognition and docking algorithms. Annual Review of Biophysics and Biomolecular Structure. 2003;**32**(1):335–373

[17] Pagadala NS, Syed K, Tuszynski J. Software for molecular docking: A review. Biophysical Reviews. 2017;**9**(2):91–102

[18] Prieto-Martínez FD, Arciniega M, Medina-Franco JL. Acoplamiento molecular: Avances recientes y retos. TIP Revista Especializada En Ciencias Químico-Biológicas. 2019;**21**(S1):65–87

[19] Shah A, Jain M. Limitations and Future Challenges of Computer-aided Drug Design Methods, in Computer Aided Drug Design (CADD): From Ligand-based Methods to Structure-based Approaches. Elsevier; 2022. 283–297

[20] Pinzi L, Rastelli G. Molecular docking: Shifting paradigms in drug discovery. International Journal of Molecular Sciences. 2019;**20**(18):4331

[21] Garg P, et al. Artificial intelligence–driven computational approaches in the development of anticancer drugs. Cancers 2024;**16**:3884

[22] Salmaso V, Moro S. Bridging molecular docking to molecular dynamics in exploring ligand-protein recognition process: An overview. Frontiers in Pharmacology. 2018;**9**:923

[23] Beichert J, et al. Innovative therapeutic strategies targeting the network architecture of glioblastoma. Clinical Cancer Research: An Official Journal of the American Association for Cancer Research. 2025;**31**(14):2864–2871

[24] D'Alessandris QG, et al. Neuromodulation for brain tumors: Myth or reality? A narrative review. International Journal of Molecular Sciences. 2023;**24**(14):11738

[25] Venkataramani V, et al. Cancer neuroscience of brain tumors: From multicellular networks to neuroscience-instructed cancer therapies. Cancer Discovery 2025;**15**:39–51

[26] Barron T, et al. GABAergic neuron-to-glioma synapses in diffuse midline gliomas. Nature. 2025;**639**:1060–1068

[27] Wang X, et al. Tumor-infiltrating nerves: Unraveling the role of cancer neuroscience in tumorigenesis, disease progression, and emerging therapies. Discover Oncology. 2025;**16**:1209

[28] Davidson B, et al. Neuromodulation techniques–From non-invasive brain stimulation to deep brain stimulation. Neurotherapeutics 2024;**21**:e00330

[29] Fu K, et al. Regulation of the brain tumor microenvironment by focused ultrasound. Molecular Therapy - Oncology 2025;**33**:200994

[30] Grenouillet S, et al. Analgesic efficacy of non-invasive neuromodulation techniques in chronic cancer pain: A systematic review. Supportive Care in Cancer 2025;**33**:1–14

[31] Krishna S, et al. Glioblastoma remodelling of human neural circuits decreases survival. Nature 2023;**617**:599–607

[32] Lv W, Wang Y. Neural influences on tumor progression within the central nervous system. CNS Neuroscience & Therapeutics. 2024;**30**(10):e70097

[33] Maier JP, et al. Inhibition of metabotropic glutamate receptor III facilitates sensitization to alkylating chemotherapeutics in glioblastoma. Cell Death & Disease 2021;**12**:723

[34] Mollazadeh H, et al. Deciphering the role of glutamate signaling in glioblastoma multiforme: Current therapeutic modalities and future directions. Current Pharmaceutical Design 2020;**26**:4777–4788

[35] Tian J, Kaufman DL. The GABA and GABA-receptor system in inflammation, anti-tumor immune responses, and COVID-19. Biomedicines. 2023;**11**(2):254

[36] Huang D, et al. GABAergic signaling beyond synapses: An emerging target for cancer therapy. Trends in Cell Biology 2023;**33**:403–412

[37] Tetzlaff SK, et al. Characterizing and targeting glioblastoma neuron-tumor networks with retrograde tracing. Cell 2025;**188**: 390–411.e36

[38] Takayasu T, et al. Ion channels and their role in the pathophysiology of gliomas. Molecular Cancer Therapeutics 2020;**19**:1959–1969

[39] Thammavongsa DA, et al. Targeting ion channels: Blockers suppress calcium signals and induce cytotoxicity across medulloblastoma cell models. Bioengineering 2025;**12**:268

[40] Bahcheli AT, et al. Pan-cancer ion transport signature reveals functional regulators of glioblastoma aggression. The EMBO Journal 2024;**43**:196–224

[41] Venkataramani V, et al. Synaptic input to brain tumors: Clinical implications. Neuro-oncology 2021;**23**:23–33

[42] Bernstock JD, et al. Gabapentinoids confer survival benefit in human glioblastoma. Nature Communications 2025;**16**:4483

[43] Radin DP, Patel P. BDNF: An oncogene or tumor suppressor? Anticancer Research. 2017;**37**(8): 3983–3990

[44] Smer-Barreto V, et al. Identification of Drug Candidates against Glioblastoma with Machine Learning and High-throughput Screening of Heterogeneous Cellular Models. bioRxiv. 2025.03.06.641926. 2025

[45] Sumera, et al. Molecular docking and molecular dynamics studies reveal secretory proteins as novel targets of temozolomide in glioblastoma multiforme. Molecules 2022;**27**:7198

[46] Sumera S, Zahid S, P121: Molecular Docking and Molecular Dynamics Studies Reveal Secretory Proteins as Novel Targets of Temozolomide in Glioblastoma Multiforme. Genetics in Medicine Open, 2025. **3**

[47] Huang S, et al. Tizoxanide promotes apoptosis in glioblastoma by inhibiting CDK1 activity. Frontiers in Pharmacology. 2022;**13**:895573

[48] Pradeep S, et al. Exploring shared therapeutic targets for Alzheimer's disease and glioblastoma using network pharmacology and protein-protein

interaction approach. Frontiers in Chemistry. 2025;**13**:1549186

[49] DeRiemer S, et al. Enhancement of calcium current in Aplysia neurones by phorbol ester and protein kinase C. Nature 1985;**313**:313–316

[50] Karaman I, Sebin B. From data-driven cities to data-driven tumors: Dynamic digital twins for adaptive oncology. Frontiers in Artificial Intelligence. 2025;**8**: 1624877

[51] Kar S, Leszczynski J. Databases for drug discovery and development. In Current Trends in Computational Modeling for Drug Discovery, Springer; 2023. 269–298

[52] Zhong S, et al. Selected by gene co-expression network and molecular docking analyses, ENMD-2076 is highly effective in glioblastoma-bearing rats. Aging (Albany NY) 2019;**11**:9738

[53] Bruzzone F, et al. Exploring the role of ADCs in brain metastases and primary brain tumors: Insight and future directions. Cancers 2025; **17**:1591

[54] Basu A, et al. An interactive resource to identify cancer genetic and lineage dependencies targeted by small molecules. Cell 2013;**154**:1151–1161

[55] Tabaee Damavandi P, et al. Perampanel in brain tumor-related epilepsy: A systematic review. Brain Sciences 2023;**13**:326

[56] Eroglu C, et al. Gabapentin receptor α2δ-1 is a neuronal thrombospondin receptor responsible for excitatory CNS synaptogenesis. Cell 2009;**139**:380–392

[57] Rzeski W, Turski L, Ikonomidou C. Glutamate antagonists limit tumor growth. Proceedings of the National Academy of Sciences. 2001;**98** (11):6372–6377

[58] Liu K-H, et al. Fluoxetine, an antidepressant, suppresses glioblastoma by evoking AMPAR-mediated calcium-dependent apoptosis. Oncotarget 2014;**6**:5088

[59] Jiang X, et al. GABAB receptor complex as a potential target for tumor therapy. Journal of Histochemistry & Cytochemistry 2012;**60**:269–279

[60] Singh N, Villoutreix BO. Demystifying the molecular basis of pyrazoloquinolinones recognition at the extracellular α1+/β3-interface of the GABAA receptor by molecular modeling. Frontiers in Pharmacology. 2020;**11**:561834

[61] Hoja S, et al. Molecular dissection of the valproic acid effects on glioma cells. Oncotarget. 2016;**7**(39):62989

[62] Sixto-López Y, Bello M, Correa-Basurto J. Exploring the inhibitory activity of valproic acid against the HDAC family using an MMGBSA approach. Journal of Computer-Aided Molecular Design. 2020;**34**(8):857–878

[63] Cox B, Ion channel drug discovery: A historical perspective. 2014.

[64] Bhrdwaj A, et al. Structure-based virtual screening, molecular docking, molecular dynamics simulation of EGFR for the clinical treatment of glioblastoma. Applied Biochemistry and Biotechnology 2023;**195**:5094–5119

[65] Kovacic P, Somanathan R. Clinical physiology and mechanism of dizocilpine (MK-801): Electron

transfer, radicals, redox metabolites and bioactivity. Oxidative Medicine and Cellular Longevity. 2010;**3**(1):13–22

[66] Misgeld U, Bijak M, Jarolimek W. A physiological role for GABAB receptors and the effects of baclofen in the mammalian central nervous system. Progress in Neurobiology. 1995;**46**(4):423–462

[67] Kim T, et al. HDAC inhibition by valproic acid induces neuroprotection and improvement of PD-like behaviors in LRRK2 R1441G transgenic mice. Experimental Neurobiology 2019;**28**:504

[68] Chen J, et al. TRPM7 channel inhibition mediates midazolam-induced proliferation loss in human malignant glioma. Tumor Biology 2016;**37**:14721–14731

[69] Mocanu MM, et al. Mibefradil, a T-type and L-type calcium channel blocker, limits infarct size through a glibenclamide-sensitive mechanism. Cardiovascular Drugs and Therapy 1999;**13**:115–122

[70] Vandenberg JI, et al. hERG K +channels: Structure, function, and clinical significance. Physiological Reviews. 2012;**92**:1393–1478

[71] Tegowski M, Fan C, Baldwin AS. Thioridazine inhibits self-renewal in breast cancer cells via DRD2-dependent STAT3 inhibition, but induces a G1 arrest independent of DRD2. Journal of Biological Chemistry. 2018;**293**(41):15977–15990

[72] Shin DJ, et al. Propofol is an allosteric agonist with multiple binding sites on concatemeric ternary GABAA receptors. Molecular Pharmacology 2018;**93**:178–189

[73] Janes PW, et al. An anti-VEGF-B antibody reduces abnormal tumor vasculature and enhances the effects of chemotherapy. Cancers 2024;**16**:1902

[74] Huang X-Y, et al. Everolimus inhibits PI3K/Akt/mTOR and NF-kB/IL-6 signaling and protects seizure-induced brain injury in rats. Journal of Chemical Neuroanatomy. 2021;**114**:101960

[75] Spiegl-Kreinecker S, et al. O 6-Methylguanine DNA methyltransferase protein expression in tumor cells predicts outcome of temozolomide therapy in glioblastoma patients. Neuro-oncology 2010;**12**:28–36

Chapter 5

Perspective Chapter: Introduction to Molecular Docking

Aastha Rana, Gagandeep Singh and Rohit Bhatia

Abstract

Molecular docking is a computational approach used to predict the binding orientation and affinity of ligands with target receptors, supporting drug discovery and biomedical research. This chapter outlines the principles, methodology, and recent advancements in docking. The process involves protein and ligand preparation, grid generation, active site prediction, and docking simulations, which collectively help identify stable complexes and optimize lead molecules. Emerging developments, including deep learning frameworks like GNINA, hybrid artificial intelligence approaches, and improved docking algorithms, have enhanced prediction accuracy and efficiency in virtual screening. Applications of molecular docking extend to drug repurposing, anticancer and antidepressant drug design, biosensor development, biomarker identification, and environmental remediation. Case studies include thymol and carvacrol derivatives as anticancer agents, chalcone derivatives with antidepressant potential, and graphene-based composites for dental restoratives. Docking also assists in evaluating interactions with metabolic enzymes, supporting drug safety and specificity. Despite limitations such as scoring accuracy and receptor flexibility, integration with molecular dynamics, machine learning, and quantum computing offers promising future directions. Molecular docking continues to be a cost-effective and versatile tool in drug development, personalized medicine, and translational research.

Keywords: molecular docking, scoring functions, case studies, enzymes, drug design

1. Introduction

The purpose of research on docking is to anticipate the desired three-dimensional arrangements [1]. The state of the art in several computational elements of virtual screening of a library of small compounds utilizing molecular docking is described in the current work [1, 2]. Docking is a popular technique for predicting how small-molecule medicinal drugs will align with their protein targets, thereby predicting the small molecule's affinity [3, 4]. Molecular modeling is a technique that creates, characterizes, and alters the structures and interactions of substances, as well as their properties, which are dependent on their three-dimensional geometries [4]. This prediction is used to virtually screen huge libraries of compounds and produces the conformation and, typically, the binding affinity of the small molecule in its projected least energy state. Sampling and scoring are the two primary steps that make up docking. A thorough search of the conformational space of the molecules being docked is referred to as sampling [5]. The ability to use the pose for lead optimization and the determination of a small molecule's proper binding pose are both necessary.

For figuring out its binding affinity, activities like virtual screening or deciding whether a molecule is significant for further experimental research, accurate assessment of binding affinity is essential [6]. Empirical scoring functions like X-Score, AutoDock Vina, and Chem-Score are used by a significant percentage of docking software [7, 8]. Using a particular method, the molecular docking program can assist us in determining the best conformation and orientation based on complementarity and pre-organization. A scoring function is then applied to forecast the binding affinity and examine the interactive mode [9]. A general docking procedure is depicted in **Figure 1**.

2. Methodology involved in molecular docking

The methodology of molecular docking involves a systematic approach to predicting the preferred orientation of a ligand when bound to a receptor (macromolecule) in order to form a stable complex. The process involves the following steps:

2.1 Step 1. Preparation of protein and ligand

The target protein's three-dimensional (3D) structure is acquired from the Protein Data Bank (PDB), a database kept up to date by the Research Collaboratory for Structural Bioinformatics (RCSB), in order to initiate the molecular docking procedure. To make sure the protein structure is appropriate for docking simulations, it must be pre-processed after downloading. The following actions are

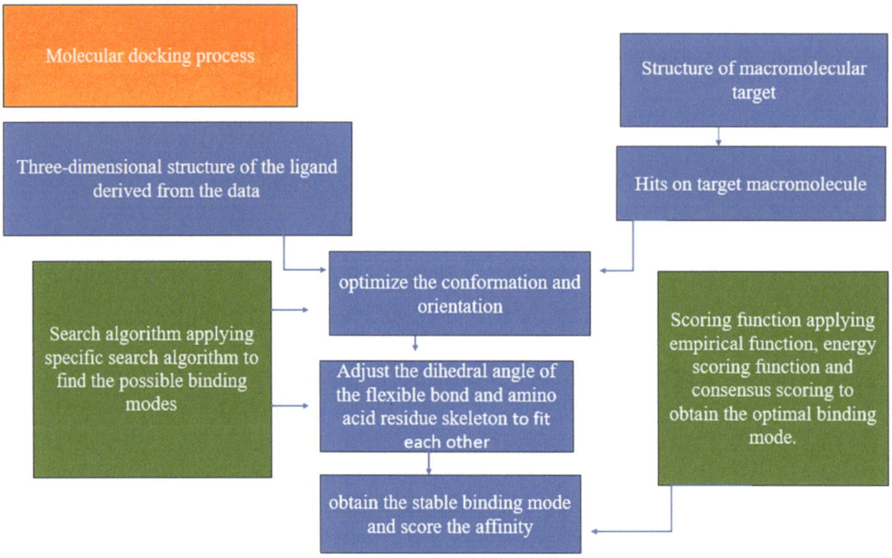

Figure 1.
Various steps involved in molecular docking.

involved in this: elimination of water molecules, which, unless they have a recognized function in the active site, may obstruct ligand binding; stabilization of charges to restore amino acid residues' proper ionization states; and completing any atoms or residues that are missing from the PDB dataset.

Hydrogen atoms, particularly polar hydrogens, are added to mimic physiological circumstances.

When required, side chains are created to finish off incomplete amino acid residues.

2.2 Step 2: Preparation of ligand

For this ligand molecule, Lipinski's Rule of Five was then applied. It is applied to both drug-like and drug-unlike substances. It lowers the failure rate and raises the high likelihood of success. Pub Chem compounds can be obtained from a variety of databases, including ZINC. It might utilize the chem sketch tool from the mol file.

2.3 Step 3: Grid generation

Each aspect is maintained in this, including the site, rotatable group, excluded volumes, and limitations. The primary determinant is the quantity of genetic operations (crossover, migration, and mutation) carried out in determining the binding cavity prediction.

2.4 Step 4: Prediction of active site

It is necessary to predict the protein molecule's active site. Following protein preparation, any water molecules and heteroatoms that may be present are removed from the cavity.

2.5 Step 5: Docking run

The interaction between ligands and proteins is examined. The optimal docking score is then determined [10–12].

3. Recent advancements in molecular docking

The following sections highlight some of the recent advances employed in molecular docking, which are making this a more promising approach in drug design and discovery.

3.1 Drug repurposing using AI, molecular docking and hybrid approaches

A research group investigated two important drug repurposing approaches – the artificial intelligence (AI) approach and the molecular docking technique – in both general disorders and Alzheimer's Disease (AD) in particular. Furthermore, the hybrid strategy that combines molecular docking methods and artificial intelligence was also investigated. In AD research, drug repurposing shows promise as a persuasive and successful tactic. In this field, molecular docking techniques and artificial intelligence both show great promise. More accurate predictions are produced by AI algorithms, which make it easier to investigate novel therapeutic uses for currently available medications. Similarly, drug–target interaction modeling is revolutionized by molecular docking approaches, which use sophisticated algorithms to evaluate large drug databases against particular target proteins. The goal is to expedite the drug discovery process, which may enhance the therapy strategy for AD [13].

3.2 GNINA1.3 the next increment in molecular docking with deep learning

This approach offers a straightforward covalent docking interface. As with previous programs, GNINA expects the bound covalent form of the ligand to be supplied as input rather than assuming a specific chemical process. The underlying deep learning framework is updated to PyTorch in this release, which makes docking more computationally efficient and opens the door for the smooth integration of additional deep learning techniques into the docking pipeline. To enable high-throughput virtual screening with GNINA, scientists introduced knowledge-distilled CNN scoring functions and retrained CNN scoring functions on the revised CrossDocked2020 v1.3 dataset [14].

3.3 Molecular docking simulation DFT computation formation of creatinine chemosensor of Cu^{2+}

Using SiO2-Pr-SO3H as a silica nano-catalyst, a multi-component reaction of isatin, malononitrile, and creatinine is carried out to create a Cu^{2+} chemosensor in a quick reaction time and with high yield, without the need for solvents. Since the Cu2 + ion is clearly one of the hazardous heavy metals for both human health and the environment, it is critical to quickly identify and detect it using an effective chemosensor. The electronic characteristics of the ligand and the ligand-Cu^{2+} complex were examined, as well as the ligand's active site for Cu^{2+} interaction, using the computational approach (DFT). The activity of the synthesized compound against the COVID-19 major protease (PDB ID: 6LU7) is examined using molecular docking modeling [15].

3.4 Modified AutoDock Vina algorithms

AutoDock Vina (Vina) has solidified its position as a top software package due to its user-friendly interface and effective energy-based scoring system. However, the growing demands of computational biology have prompted efforts to improve Vina, leading to several algorithmic breakthroughs. This in-depth assessment examines Vina's most recent developments in molecular docking. These changes include novel scoring functions combined with machine learning and hybrid parallelization techniques that utilize high-performance computing [16].

4. Softwares used for molecular docking

Some of the commonly used softwares which are easily accessible, have been depicted in **Figure 2**.

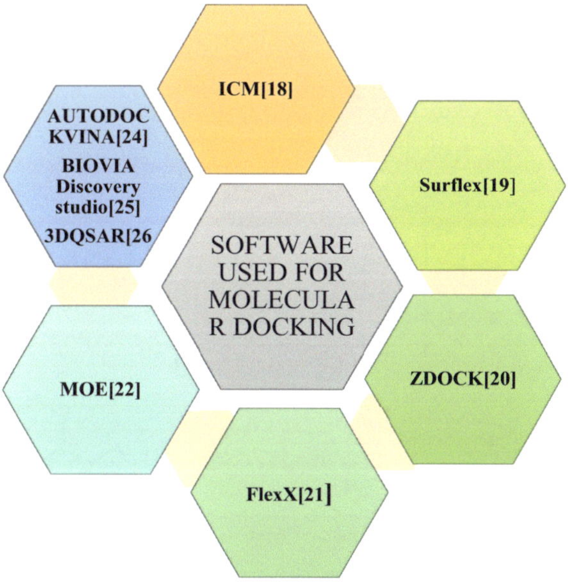

Figure 2.
Some commonly used softwares for molecular docking.

5. Recent applications of molecular docking

1. **Design of thymol and carvacrol derivatives against cancer**
   By the help of computational analysis of thymol and carvacrol derivatives as anticancer agents, several thymol and carvacrol compounds have been created. To discover and evaluate the possible biological targets associated with cancer, computational techniques such as network pharmacology and molecular docking approaches were employed. The ethoxy-cyclohexyl analogues were consistently the most effective among the produced compounds against a panel of 10 distinct cancer cell lines from various origins (see **Figure 3**). The AKT1 protein was identified as a core and central target of the most active drugs based on biological target predictions. This protein has a binding pocket that the most potent substances bind to, according to molecular docking [17, 18].

2. **Design of chalcone derivatives connected to N-heterocyclic compounds**
   New chalcone derivatives with a benzo[d][1–3]triazol-1-yl substituent were designed using a molecular docking approach, suggesting increased antidepressant efficacy (see **Figure 4**). The activity of chalcone compounds containing morpholino and piperidin-1-yl groups was comparatively reduced. The chalcone derivatives demonstrated excellent binding affinity with a good docking score [19].

3. **Design of graphene-based analogs**
   A molecular docking and dynamics investigation of graphene and its modifications for improved adherence in dental restoratives was reported

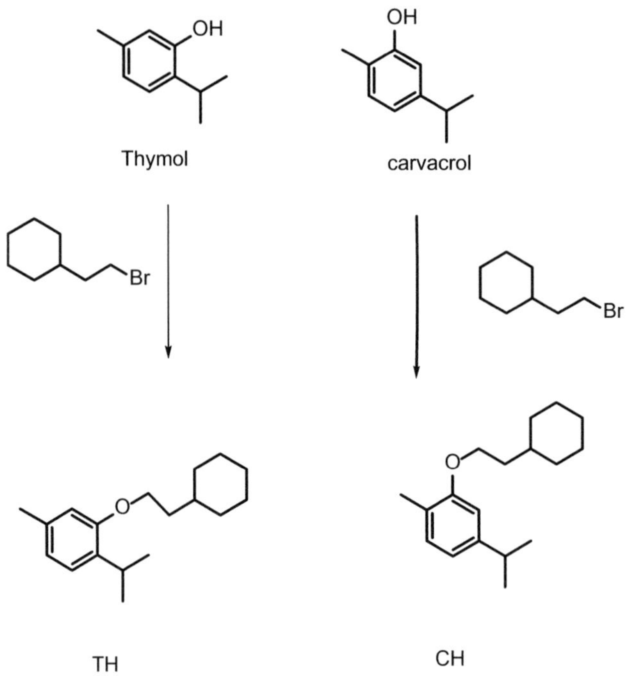

Figure 3.
Structure of designed thymol (TH) and carvacrol (CH) analogs.

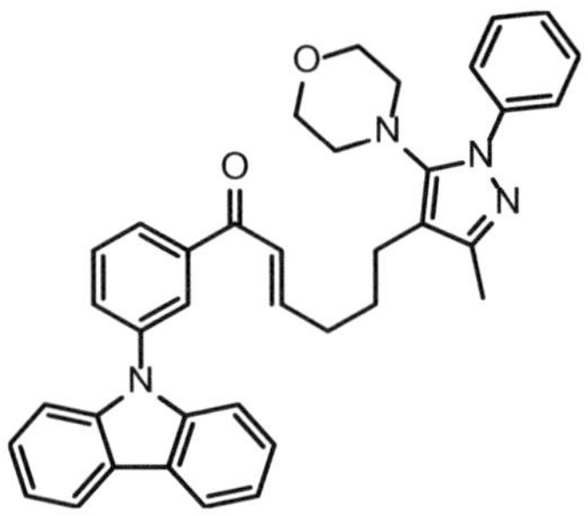

Figure 4.
Structure of designed potent chalcone derivatives.

recently. Because of its structural and adhesive qualities, which have improved the mechanical performance of dental composites, graphene has garnered a lot of interest in the dental field. For this, molecular docking and molecular dynamics (MD) simulations were adopted to study the behavior and interactions of monomers and graphene-based adhesives. Molecular docking was used to evaluate the binding energies and interactions between monomers and graphene derivatives (see **Figure 5**), as well as MD simulations [20–23].

4. **Discovery of significant biomarker genes for the occurrence of kidney renal clear cell carcinoma and liver hepatocellular carcinoma**
   A combined high-throughput screening and molecular docking

strategy, with potential for specialized treatment interventions, was developed for the discovery of cancer biomarkers. The cancer-fighting capabilities of andrographolide, particularly against liver hepatocellular carcinoma (LIHC; see **Figure 6**), have been emphasized by molecular docking and network pharmacology [24, 25].

5. **Remediation of materials**
   Protein-ligand docking can also be used to predict which pollutants are enzyme-degradable. The most effective drug may be identified using it, as well as its desired location. Enzymes and their mechanisms of action can also be determined using molecular docking. It can also be applied to ascertain the connections among proteins. The remediation method is used to virtually filter molecules [26].

6. **Molecular modeling in modern drug development**
   Molecular docking is employed to assess any damage caused by interaction with other proteins, including cytochrome P450 and proteases. Docking is used to ascertain proposed drug specificity against homologous proteins, and understanding cellular relationships facilitates the discovery of possible pharmacological targets [27].

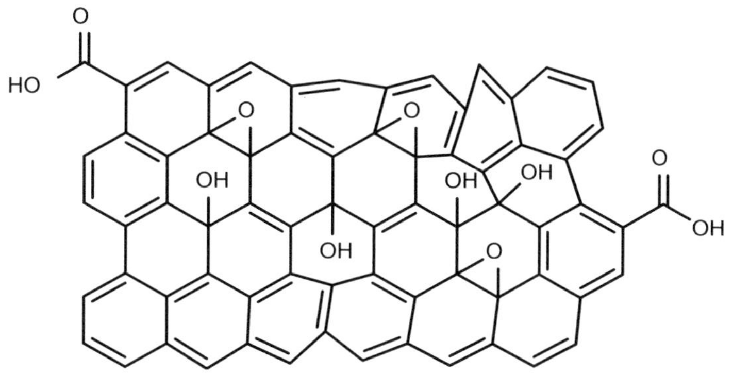

Figure 5.
Structure of graphene oxide.

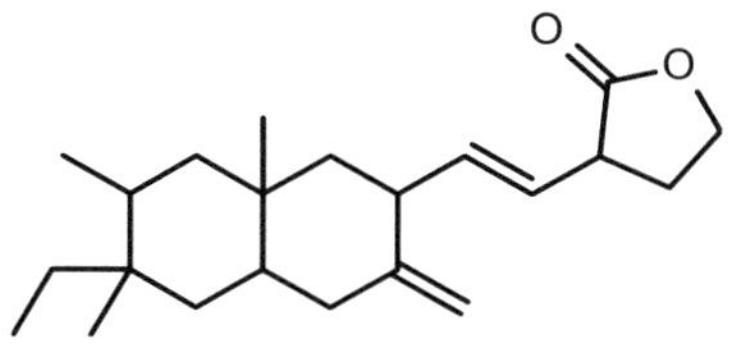

Figure 6.
Structure of the identified carcinoma biomarker.

6. Conclusion and future perspectives

Numerous tools used for drug design and discovery are available through molecular docking. Medical chemists make it easy to view structured molecular databases. It predicts ligand-receptor attachments with accuracy. It is employed in the creation of new drugs and to learn about the novel drug development and design process. The complications of the molecular docking method include lead molecule optimization, biological pathway assessment, and *de novo* drug design [28–31]. With advances in MD, artificial intelligence, machine learning, and quantum computing, computational docking in drug development should have a bright future with prompt and relevant results. These contemporary techniques, such as drug repurposing and hybrid approaches, have improved drug discovery and adhered to repurposing workflows [32]. By adjusting medication selection based on each patient's unique genetic profile, molecular docking is expected to play a significant role in personalized medicine and enhance therapeutic outcomes. High-throughput virtual screening pipelines and cloud-based docking platforms will further democratize drug discovery resources.

Author details

Aastha Rana, Gagandeep Singh and Rohit Bhatia*
Chitkara College of Pharmacy, Chitkara University, Rajpura, India

*Address all correspondence to: bhatiarohit5678@gmail.com

References

[1] Raval K, Ganatra T. Basics, types and applications of molecular docking: A review. IP International Journal of Comprehensive and Advanced Pharmacology. 2022;**7**(1):12–16

[2] Morris GM, Lim-Wilby M. Molecular docking. Methods in Molecular Biology. 2008;**443**:365–382. DOI: 10.1007/978-1-59745-177-2_19

[3] Huang N, Shoichet BK, Irwin JJ. Benchmarking sets for Molecular Docking. Journal of Medicinal Chemistry. 2006;**49**(23):6789–6801. DOI: 10.1021/jm0608356

[4] Shoichet BK, Mcgovern SL, Wei B, Irwin JJ. Lead discovery using molecular docking. Current Opinion in Chemical Biology. 2002;**6**(4):439–485. DOI: 10.1016/s1367-5931(02)00339-3

[5] Kitchen DB, Decornez H, Furr JR, Bajorath J. Docking and scoring in virtual screening for drug discovery: Methods and applications. Nature Reviews Drug Discovery. 2004;**3**(11):935–949

[6] McNutt AT, Francoeur P, Aggarwal R, Masuda T, Meli R, Ragoza M, Sunseri J, Koes DR. GNINA 1.0: Molecular docking with deep learning. Journal of Cheminformatics. 2021;**13**(1):43

[7] Wang R, Lai L, Wang S. Further development and validation of empirical scoring functions for structure-based binding affinity predic tion. Journal of Computer-Aided Molecular Design. 2002;**16**(1):11–26

[8] Trott O, Olson AJ. Autodock vina: Improving the speed and accu racy of docking with a new scoring function, efficient optimization, and multithreading. Journal of Computational Chemistry. 2010;**31**(2):455–461

[9] Fan J, Fu A, Zhang L. Progress in molecular docking. Quantitative Biology. 2019;**7**:83–89

[10] Bhagat RT, Butle SR, Khobragade DS, Wankhede SB, Prasad CC, Mahure DS, Armarkar AV. Molecular docking in drug discovery. Journal of Pharmaceutical Research International. 2021;**33**(30):46–58

[11] Guimaraes CR, Mathiowetz AM. Addressing limitations with the MM-GB/SA scoring procedure using the WaterMap method and free energy perturbation calculations. Journal of Chemical Information and Modeling. 2010;**50**(4):547–559

[12] Singh N, Warshel A. Absolute binding free energy calculations: On the accuracy of computational scoring of protein-ligand interactions. Proteins: Structure, Function, and Bioinformatics. 2010;**78**(7):1705–1723

[13] NA Z, TC T, AV C, Sw C. Drug Repurposing using Artificial Intelligence, Molecular Docking, and Hybrid Approaches: A Comprehensive Review in General Diseases vs Alzheimer's Disease. Computer Methods and Programs in Biomedicine. 2025;**13**:108604

[14] McNutt AT, Li Y, Meli R, Aggarwal R, Koes DR. GNINA 1.3: The next increment in molecular docking with deep learning. Journal of Cheminformatics. 2025;**17**(1):28

[15] Jin Z, Du X, Xu Y, Deng Y, Liu M, Zhao Y, Zhang B, Li X, Zhang L, Peng C, Duan Y. Structure of Mpro from SARS-CoV-2 and discovery of its inhibitors. Nature. 2020;**582**(7811): 289–293

[16] Sarkar A, Concilio S, Sessa L, Marrafino F, Piotto S. Advancements and novel approaches in modified AutoDock Vina algorithms for enhanced molecular docking. Results in Chemistry. 2024;**7**:101319

[17] Thymol and carvacrol derivatives as anticancer agents; synthesis, in vitro activity, and computational analysis of biological targets†

[18] Dallakyan S, Olson AJ. Small-molecule library screening by docking with PyRx, Methods Mol. Biology. 2015;**1263**:243–250

[19] Deshmukh HS, Adole VA, Wagh SB, Khedkar VM, Jagdale BS. Exploring N-heterocyclic linked novel hybrid chalcone derivatives: Synthesis, characterization, evaluation of antidepressant activity, toxicityassessment, molecular docking, DFT and ADME study. RSC Advances. 2025;**15**(20):16187–16210

[20] Saghiri MA, Saini RS, Kuruniyan MS, Mosaddad SA, Heboyan A. Graphene and its modifications for enhanced adhesion in dental restoratives: A molecular docking and dynamics study. Scientific Reports. 2025;**15**(1):9455

[21] Cousins KR. Computer review of ChemDraw ultra 12.0. Journal of the American Chemical Society. 2011;**133** (21):8388

[22] Dominguez C, Boelens R, Bonvin AMJJ. HADDOCK: A Protein–Protein Docking approach based on biochemical or biophysical information. Journal of the American Chemical Society. 2003;**125**(7): 1731–1737

[23] Vangone A, Bonvin AMJJ. PRODIGY: A Contact-based predictor of binding affinity in Protein-protein complexes. Bio-protocol. 2017;**7**(3): e2124. DOI: 10.21769/BioProtoc.2124

[24] Dixit N, et al. Decoding the mechanism of andrographolide to combat hepatocellular carcinoma: A network pharmacology integrated molecular docking and dynamics approach. Journal of Biomolecular Structure and Dynamics. 2023;**42**(19):10237–55

[25] Tasneem M, Gupta SD, Jony MJ, Minkara M, Dey RK, Ferdoush J. Identification of key biomarker genes in liver hepatocellular carcinoma and kidney renal clear cell carcinoma progression: A shared high-throughput screening and molecular docking method with potentials for targeted therapeutic interventions. Journal of Genetic Engineering and Biotechnology. 2025;**23**(2):100497

[26] Gschwend DA, Good AC, Kuntz ID. Molecular docking towards drug discovery. Journal of Molecular Recognition. 1996;**9**(2):175–186

[27] Pagadala NS, Syed K, Tuszynski J. Software for molecular docking: A review. Biophysical Reviews. 2017;**9** (2):91–102

[28] Mitchell JBO, Laskowski RA, Alex A, Thornton JM. Bleep-potential of mean force describing protein-ligand interactions: I. generating potential. Journal of Computational Chemistry. 1999;**20** (11):1165–1176

[29] Ishchenko AV, Shakhnovich EI. Small Molecule Growth 2001 (SMoG2001): Animproved knowledge-based scoring function for protein-ligand interactions. Journal of Medicinal Chemistry. 2002;**45** (13):2770–2780

[30] Feher M, Deretey E, Roy S. BHB: A simple knowledge-based scoring function to improve the efficiency of database screening. Journal of Chemical Information and Computer Sciences. 2003;**43**(4):1316–1327

[31] Verkhivker G, Appelt K, Freer ST, Villafranca JE. Empirical free energy calculations of ligand-protein crystallographic complexes. I. Knowledge based ligand-protein interaction potentials applied to the prediction of human immunodeficiency virus 1 protease binding affinity. Protein Engineering, Design & Selection. 1995;**8** (7):677–691

[32] Wallqvist A, Jernigan RL, Covell DG. A preference-based free-energy parameterization of enzyme-inhibitor binding. Applications to HIV-1-protease inhibitor design. Protein Science. 1995;**4**(9):1881–1903